IMAGES
*of America*

# BLACK RIVER CANAL

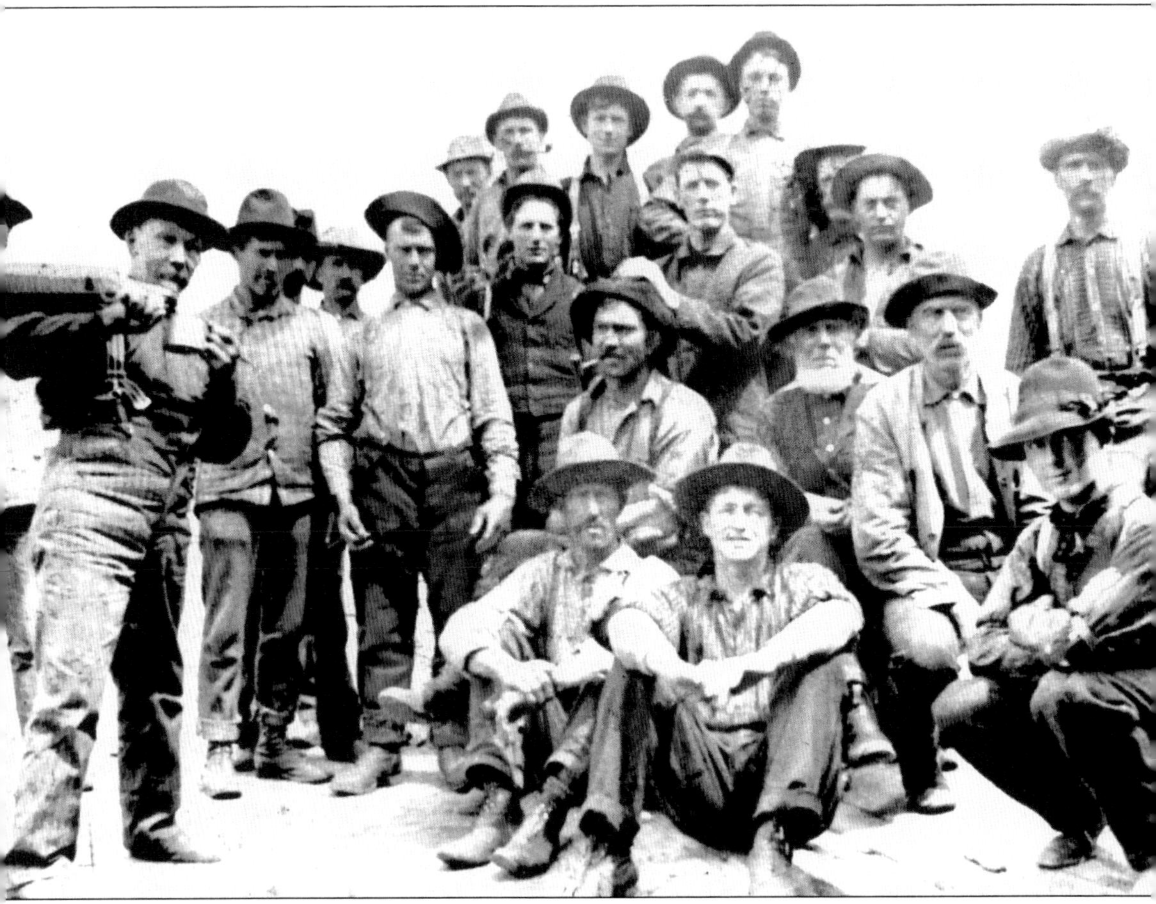

State workers kept the canal functioning. They repaired locks, removed debris, and fixed breaks in the banks.

IMAGES
*of America*

# BLACK RIVER CANAL

Edward P. Fynmore and Harney J. Corwin

ARCADIA
PUBLISHING

ISBN 978-0-7385-3813-6

Published by Arcadia Publishing
Charleston, South Carolina

Printed in the United States of America

Library of Congress Catalog Card Number: 2005922599

For all general information, contact Arcadia Publishing:
Telephone 843-853-2070
Fax 843-853-0044
E-mail sales@arcadiapublishing.com
For customer service and orders:
Toll-free 1-888-313-2665

Visit us on the Internet at www.arcadiapublishing.com

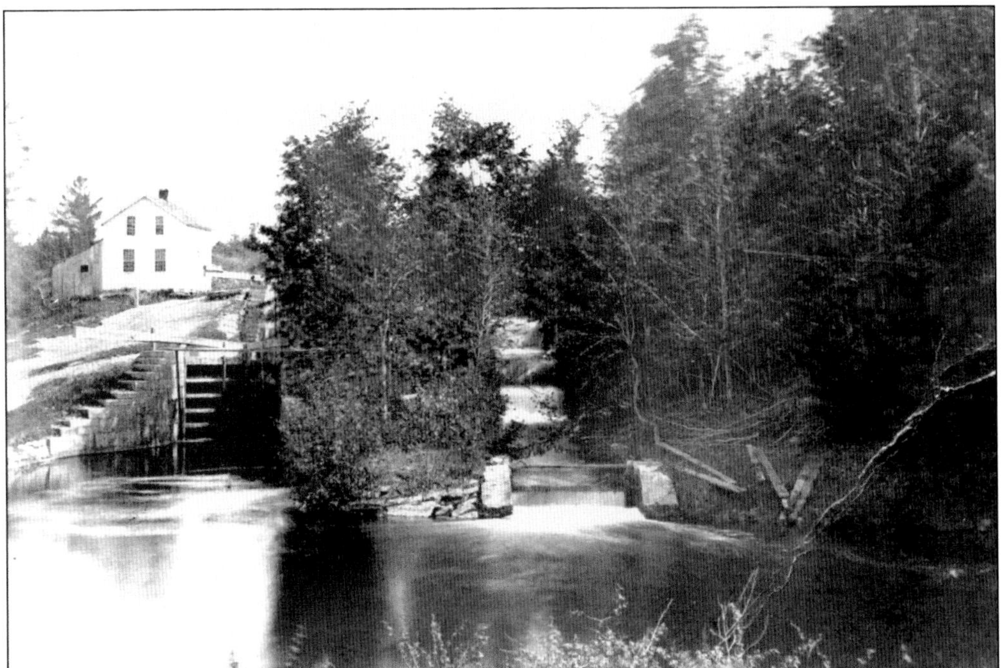

Lock 109 in Lyons Falls marked the end of the overland Black River Canal and the beginning of the canalized Black River. (Courtesy of Mary Teal.)

# Contents

# ACKNOWLEDGMENTS

We would like to thank Craig Williams of the New York State Museum, who guided our research and gave us access to Earl Scouthon's collection of images. We would like to extend thanks to Andy Kitzmann for use of photographs in the Erie Canal Museum's collection. Our gratitude goes out to the following organizations: Carthage Free Library, Rome Historical Society, Lewis County Historical Society, Port Leyden Community Library, and Town of Webb Historical Association. And we would like to thank the many individuals who generously shared their personal collections with us, including George Cataldo, Mary Centro, Ken Clark, George Davis, Vincent T. Fitch, Paul Joslyn, Bob Hindman, Joann Sattler, Vivian Seeloff, Mary Teal, and Todd Weseloh.

# Introduction

With the Erie Canal completed in 1825, Gov. DeWitt Clinton proposed construction of a northern canal that would link the Erie with Lake Ontario. This canal was to make use of the Black River, which rises in the central Adirondacks and flows north, dividing the Tug Hill Plateau on the west from the Adirondacks on the east. Clinton argued that a Black River canal would serve two purposes. First, it would open up the sparsely populated North Country to commerce. It was widely recognized that this isolated region of New York possessed vast forests, iron deposits, and fertile lands ready for exploitation. Second, a Black River canal would tap the region's abundant rivers and lakes. A reliable source of water would be essential for any expansion of the Erie Canal.

In 1836, the state legislature authorized work on the Black River Canal. Although the proposed canal would only be 35 miles long, its construction presented engineers with a daunting challenge. The channel was to proceed north from Rome along the Mohawk River to the junction of the Lansing Kill. The canal was then to proceed north through the rugged Lansing Kill Gorge, rising 693 feet to its summit at Boonville. This so-called alpine section of the canal would prove the most difficult. The canal had to be carved out of shale. This soft, brittle rock not only holds water poorly but also lends itself to the formation of sinkholes that might cause a wall or even the bottom of the canal to disappear. In places, the towpath, which formed one wall of the canal, had to be erected on precipitous hillsides high above the Lansing Kill. Here, freshets threatened to wash out the canal and landslides threatened to bury it. At other places, where the gorge was narrow, the canal and Lansing Kill ran side by side, separated only by the towpath. From Boonville, construction was somewhat less problematic. The canal would continue north to Lyons Falls on the Black River, descending 386 feet over a foundation of limestone and metamorphic rock. From Lyons Falls to its terminus at Carthage, a distance of some 42 miles, the Black River itself was to be improved for slack-water navigation. Beyond Carthage, the river was too rough for large boats.

Contracts for the entire canal had been let by 1840, but work was suspended between 1842 and 1847 when the legislature passed a Stop and Tax Law. Nonetheless, in the spring of 1850, the southern section of the canal between Rome and Boonville opened. In the following year, the canal was open as far as Port Leyden, and by 1855, the entire canal was operational. The Black River Canal was an engineering marvel. Though only 35 miles long, it required 109 locks to negotiate a rise and fall of 1,079 feet. The much longer Erie Canal stretched 363 miles but required only 83 locks over a rise and fall of 692 feet. Between Rome and Boonville—a distance of 25 miles—70 locks were built, and between Boonville and Lyons Falls—a distance of 10 miles—another 39 were constructed. Built largely by Irish immigrants, who were paid a paltry $17.50 for a quarter-year's work, the locks were fashioned out of locally available limestone, much of it quarried at nearby Sugar River. Many of the locks are still extant today, including a five-lock flight in the Lansing Kill Gorge and two four-lock combinations just north of Boonville. Besides the locks, various dams, aqueducts, bridges, bulkheads, waste weirs, sluices, culverts, and lock houses had to be built.

As construction on the overland canal went forward, the Black River north of Lyons Falls was canalized—deepened and stabilized with dams, jetties, pilings, and two more locks. Whereas draft animals towed boats on the overland canal, side-wheel steamers were used to tow barges on the Black River. The completed canal cost $3.5 million, $1 million above the original estimate and nearly half of what the Erie Canal had cost. The final cost included a 10-mile feeder between Forestport and Boonville, which would supply water to both the Black River and Erie Canals. When this source of water proved inadequate, the state built an extensive impoundment system, damming nearly every river and lake in the region. This system had a holding capacity of four billion cubic feet of water.

The Black River Canal brought a wave of prosperity to the North Country. Canal towns suddenly had busy waterfronts with new businesses springing up to meet canallers' needs. Boatbuilders and shipping agents were in demand. The state hired lock tenders, and it recruited maintenance crews to keep the canal in repair. For the first time, the vast forests of the western Adirondacks were open to logging. Axe-wielding woodsmen invaded the wilderness, and sawmills sprang up throughout the region. In 1851, the canal carried 23,000 tons in freight. By 1889, it carried 143,000 tons, and more than 90 percent of that tonnage was timber and wood products. When land was set aside for the Adirondack Park in 1892, at least two-thirds of the area had been lightly logged at least once. As local forests were cleared, farming became an increasingly important industry. Cheese and butter were shipped on the canal, as were wheat, rye, maize, and root vegetables. Over time, potatoes became the dominant export crop, although that crop never rivaled the export of timber. The new prosperity was evident in the architecture of canal towns. In Boonville, handsome new stores and striking Italianate residences graced the streets. As the great northern forest disappeared, however, canal traffic dwindled, and the Black River Canal closed in 1922. But, before it did so, it had altered the physical and economic landscape of an entire region.

# *One*

# SOUTHERN CANAL

In 1836, the New York state legislature, with Gov. William Marcy's blessing, authorized construction of the Black River Canal—but the legislature did not specify how engineers were to build an uphill canal through the rugged Lansing Kill Gorge.

The similarly steep Morris Canal in New Jersey had employed hydraulic-inclined planes. After surveying the region in 1837, chief engineer Porteous Root decided that large boats like those plying the Erie Canal required locks. In fact, Root calculated, the southern section of the canal from Rome to Boonville would require 70 locks.

By the end of 1838, contracts had been let for most of the canal. Irish shantytowns appeared, and work proceeded apace until 1842, when the legislature, controlled by the Democrats, passed a Stop and Tax Law to embarrass the Whigs, who wanted to finance all public works in the state through bonds rather than taxes. All construction except necessary repairs was suspended, and in 1844, the canal commissioners were ordered to sell off construction materials and apply the proceeds to maintenance and repairs. Fortunately, much construction had been already completed when work resumed in 1847. In 1849, the canal was filled and the first boat from Rome reached Boonville on May 10, 1850.

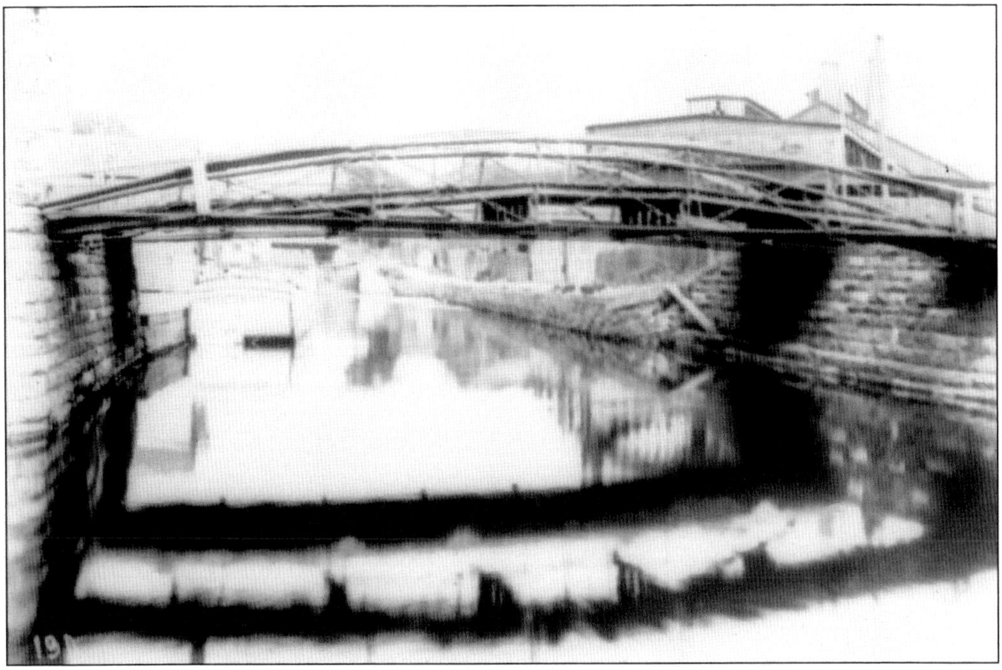

The Whitesboro Street bridge stood just north of the junction of the Black River and Erie Canals. Bridges also crossed the Black River Canal at Dominick, Stanwix, and Thomas Streets in Rome. Lock 1 is in the distance. (Courtesy of New York State Museum.)

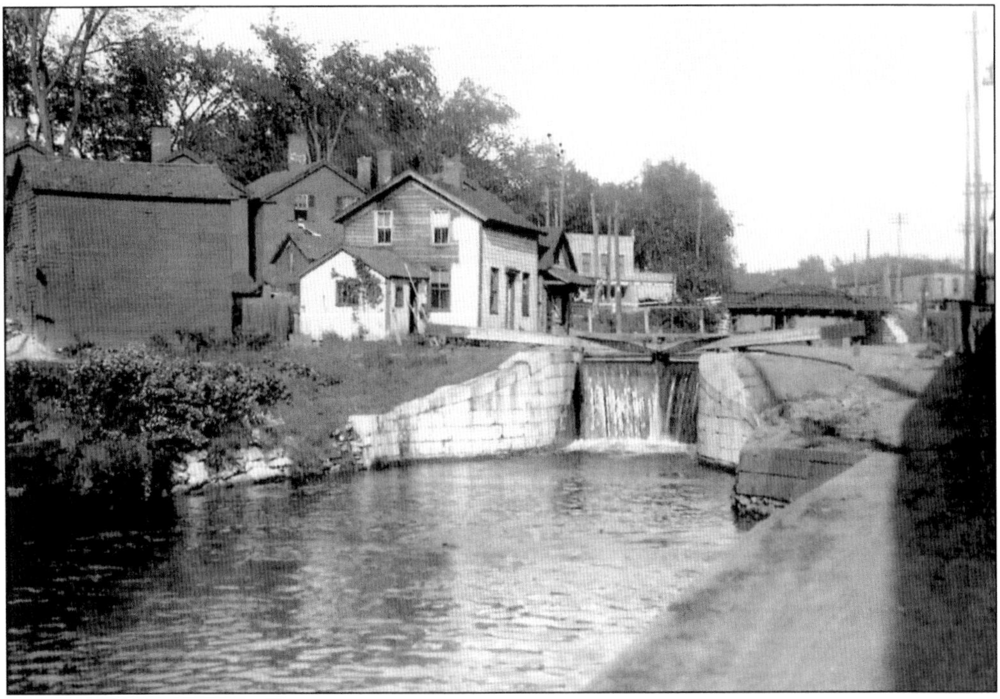

Lock 1 was 600 feet north of the Erie Canal. A sluiceway on the berm side of each lock (not visible here) drained off excess water, but water frequently spilled over closed gates. (From Rome Historical Society's Photographic Collection.)

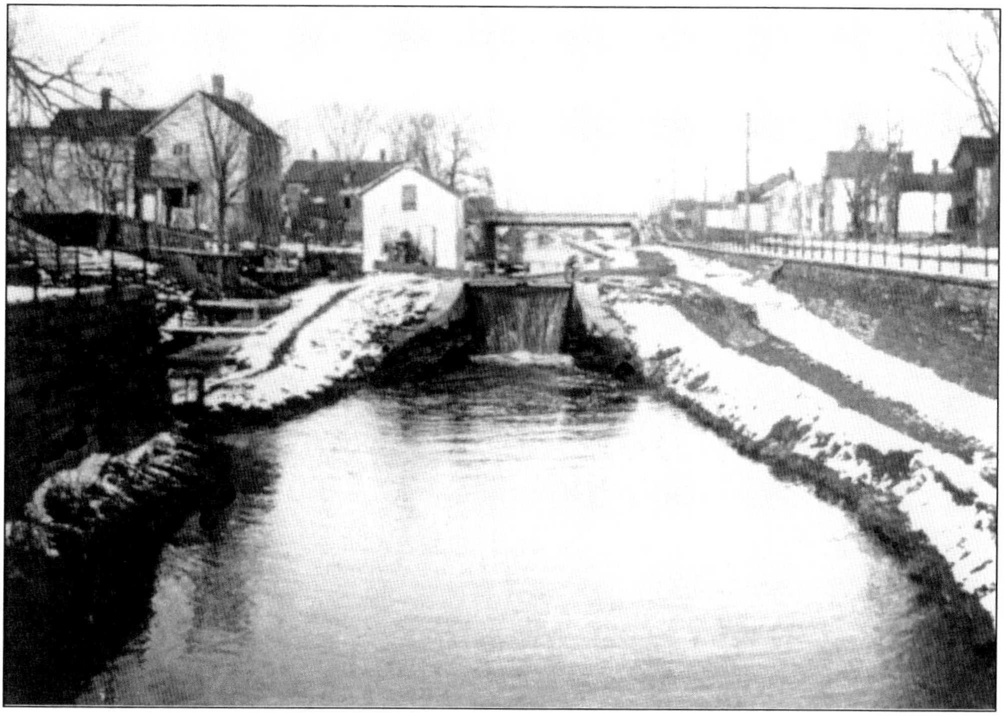

Snow is on the ground, but water is running through Lock 3 in Rome. The shipping season typically lasted from May 1 to December 1. Unexpected cold snaps, however, could strand boatmen and their families for the duration of the winter.

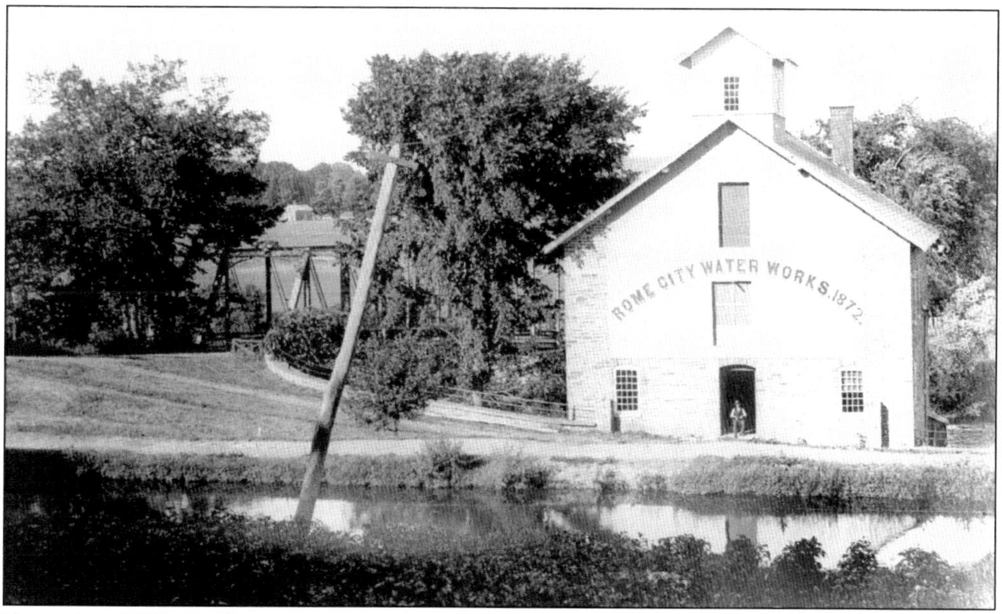


Erected in 1872, the Rome Water Works at Ridge Mills was a notable landmark on the canal. (Courtesy of New York State Museum.)

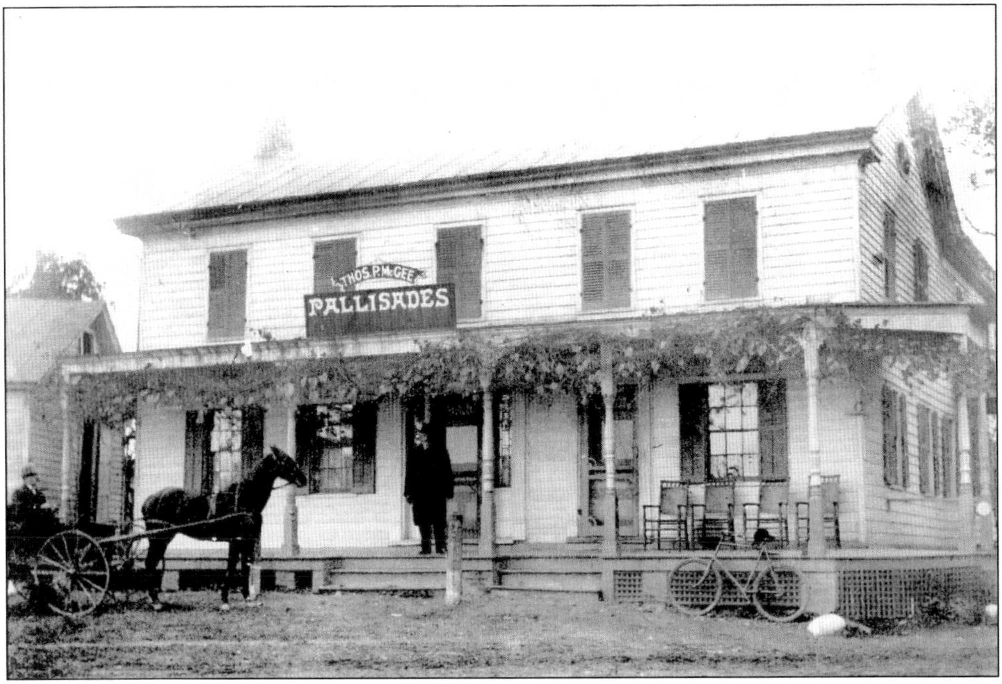


At Ridge Mills, Thomas P. McGee ran the Pallisades Hotel, named for the nearby gorge carved out by the Mohawk River. By the 1850s, boat captains in New York were generally regarded as an abstemious and sober lot. The hoggees, who drove the draft animals, developed a very different reputation, however. (Courtesy of Mary Centro.)

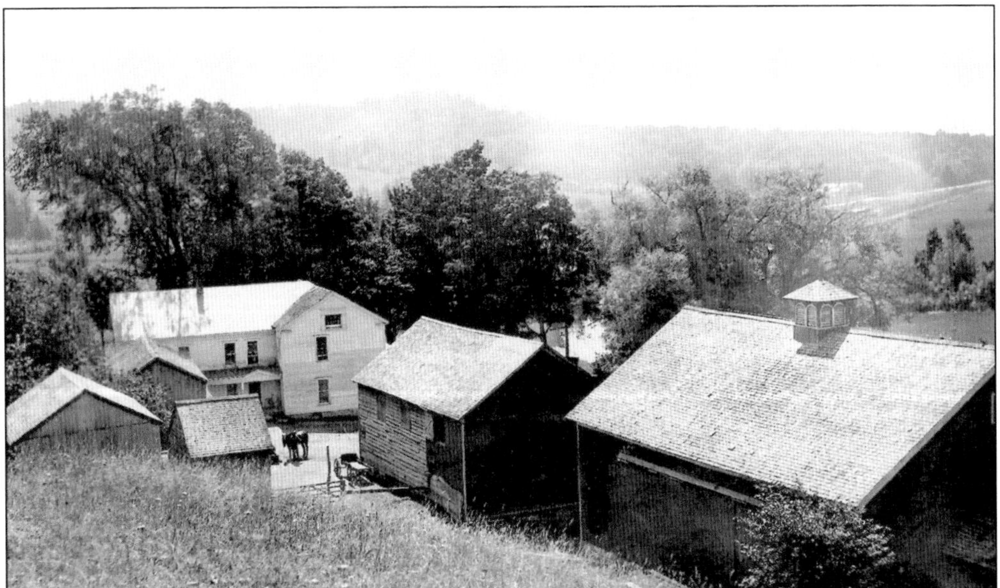

In 1851, Jesse Williams of Rome founded the first cheese factory in America, processing whole milk instead of just curds. The factory was near the Black River Canal, where the state fish hatchery is now located. (Courtesy of Mary Centro.)

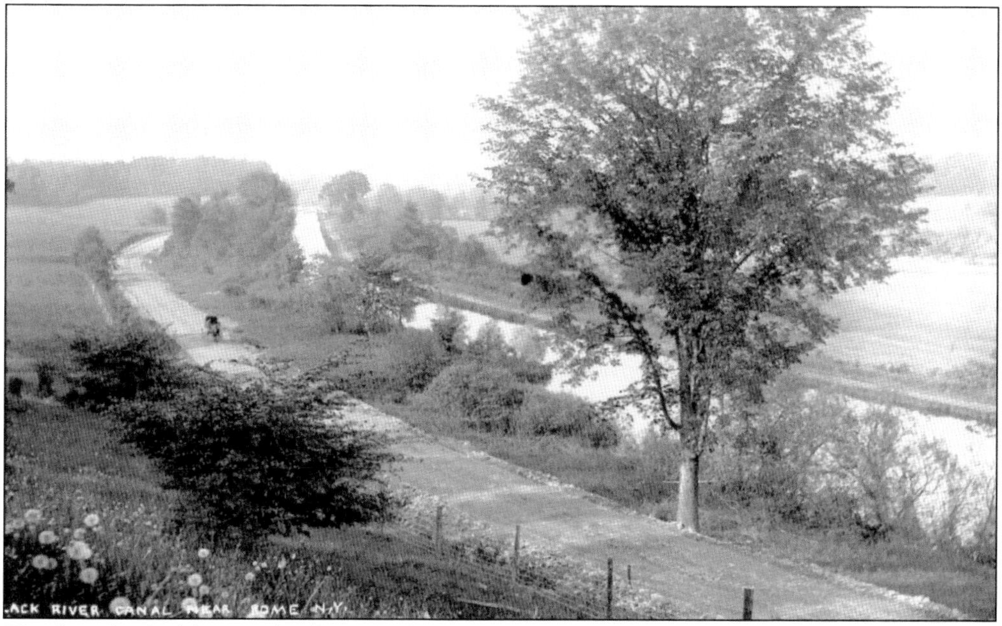


The first third of the overland canal—between the Erie Canal and Frenchville—was relatively flat, with only 17 of the canal's 109 locks. (Todd Weseloh Collection.)

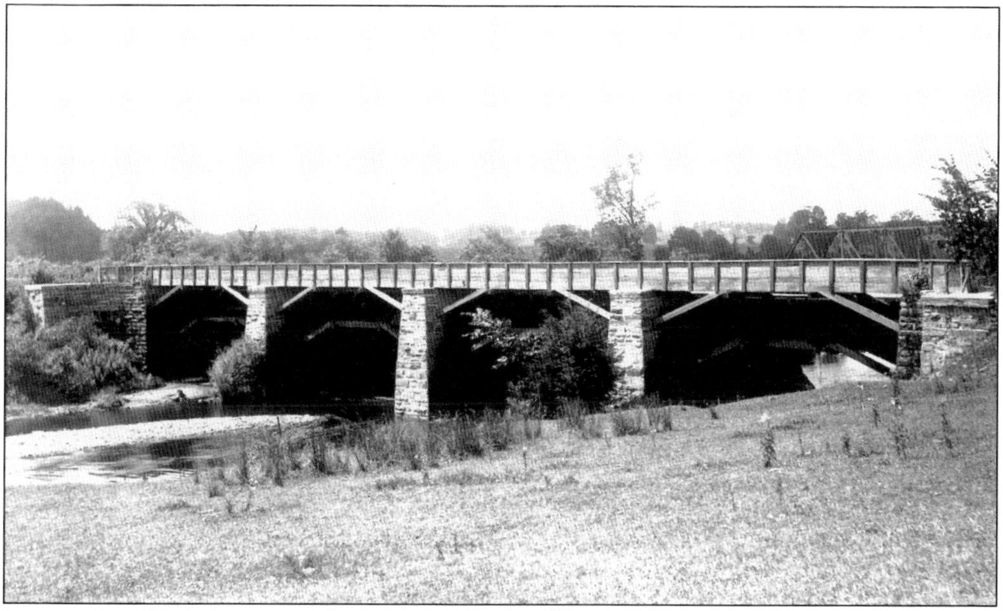

Just north of Lock 8, an aqueduct carried the canal across the Mohawk River. The structure had four 40-foot wooden arches supported by 5-foot-wide stone piers. On hot summer days, local children stood beneath the aqueduct and cooled off in the spray drifting down from above. Other aqueducts on the southern section of the canal were at Frenchville (Wells Brook-Big Brook), North Western (Stringer Creek), and Dunn Brook (Lansing Kill). (Courtesy of Erie Canal Museum.)

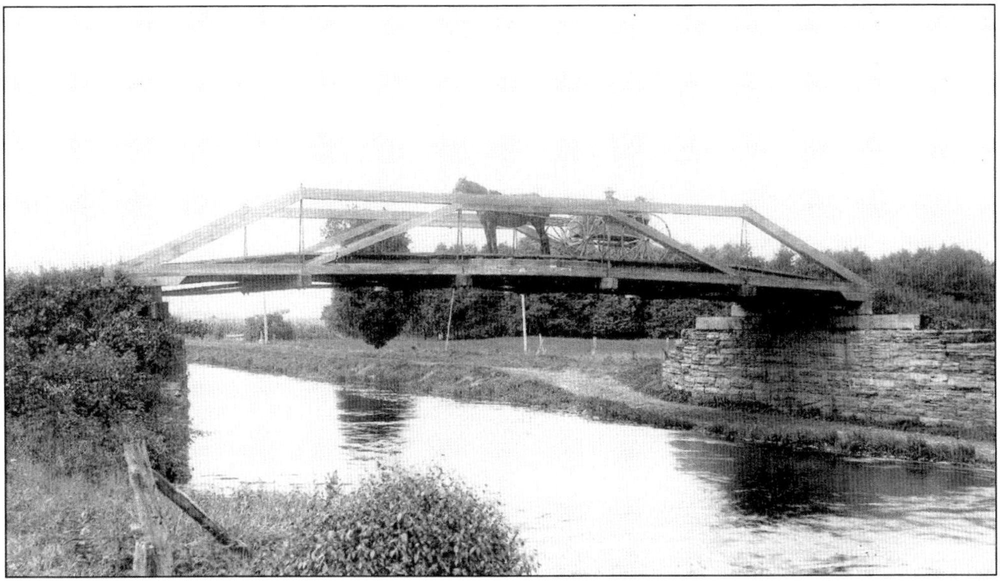

Completed in 1860, the Delta Feeder connected the Mohawk River with the Black River Canal 75 feet north of the Mohawk aqueduct. In 1892, five bridges crossed the Delta Feeder. (Courtesy of Erie Canal Museum.)

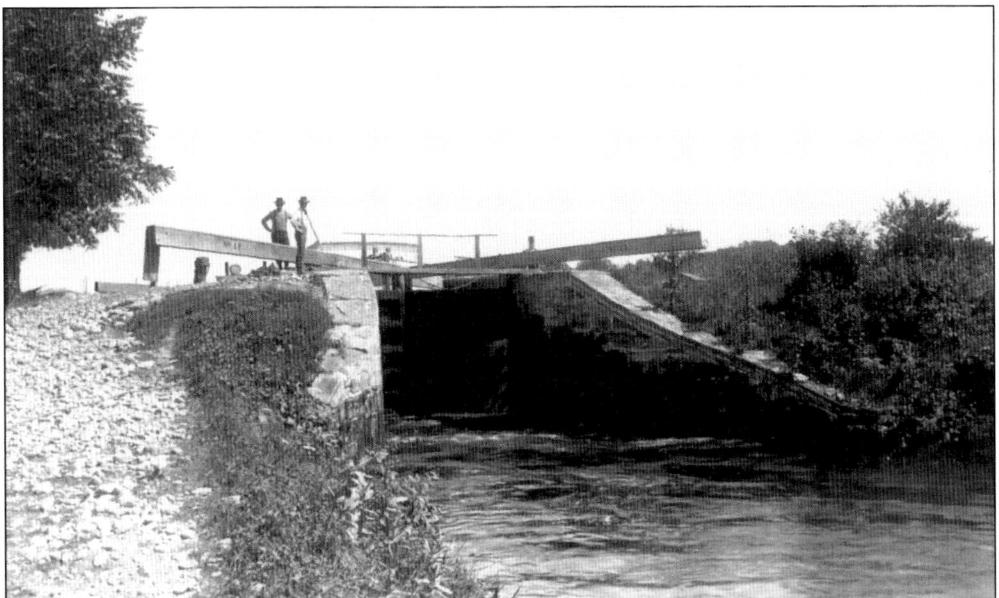

A boat heads north through Lock 11, six miles north of the Erie Canal, near the village of Delta. (Courtesy of Erie Canal Museum.)

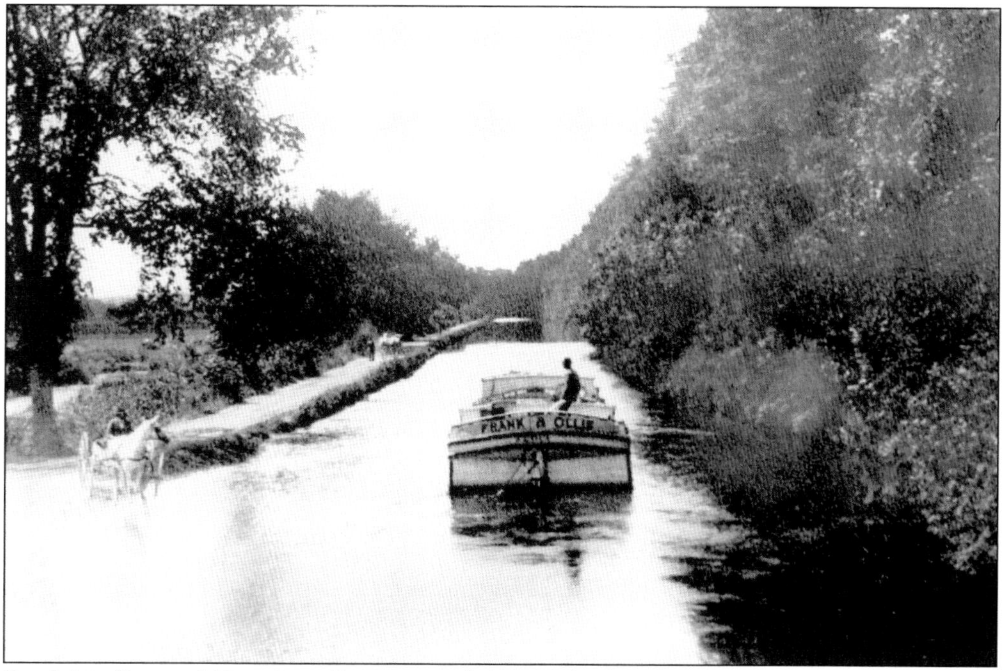


Frank Hurlbut and Tom Carroll, captain of the *Frank and Ollie*, exchange gossip near Delta. The typical Black River Canal boat was about 75 feet long and could carry 70 tons, or about 3,000 bushels of potatoes. A fully loaded boat had a draft of 3.5 feet.

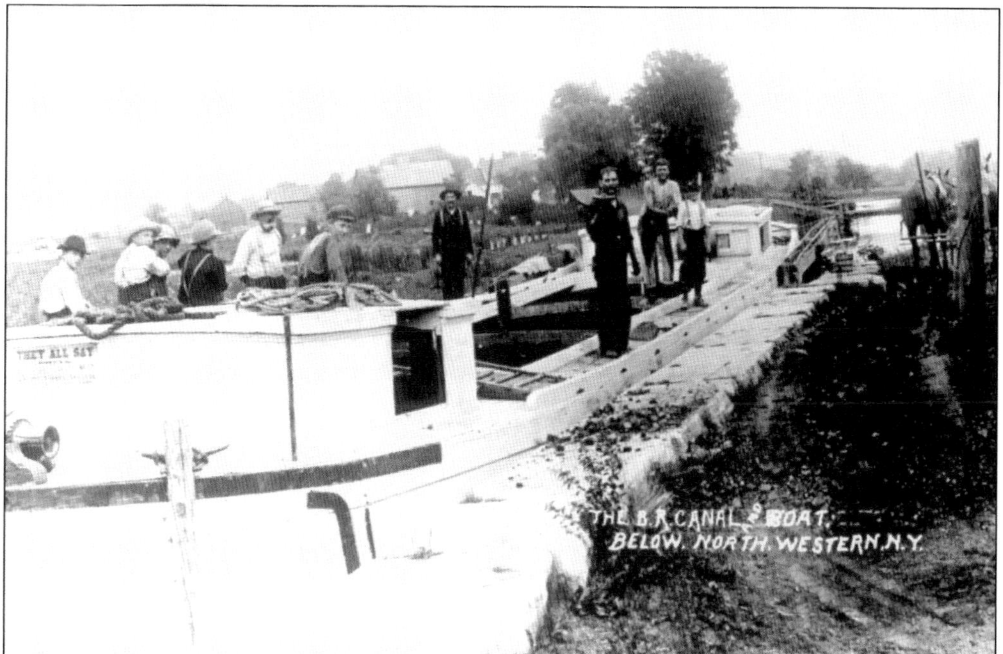


Boats were built to fit snugly in locks measuring 90 feet by 15 feet. The people aboard this boat are, from left to right, Frank O'Dell, Fred Beckley, Charles Beckley, Virgil Barnard, Thomas Carroll, William Nutt, Will Martin, Henry Smith, Bert Vandewalker, and Clinton Moore. In the background is the Edward Rudd farm and cemetery at Delta, not North Western.

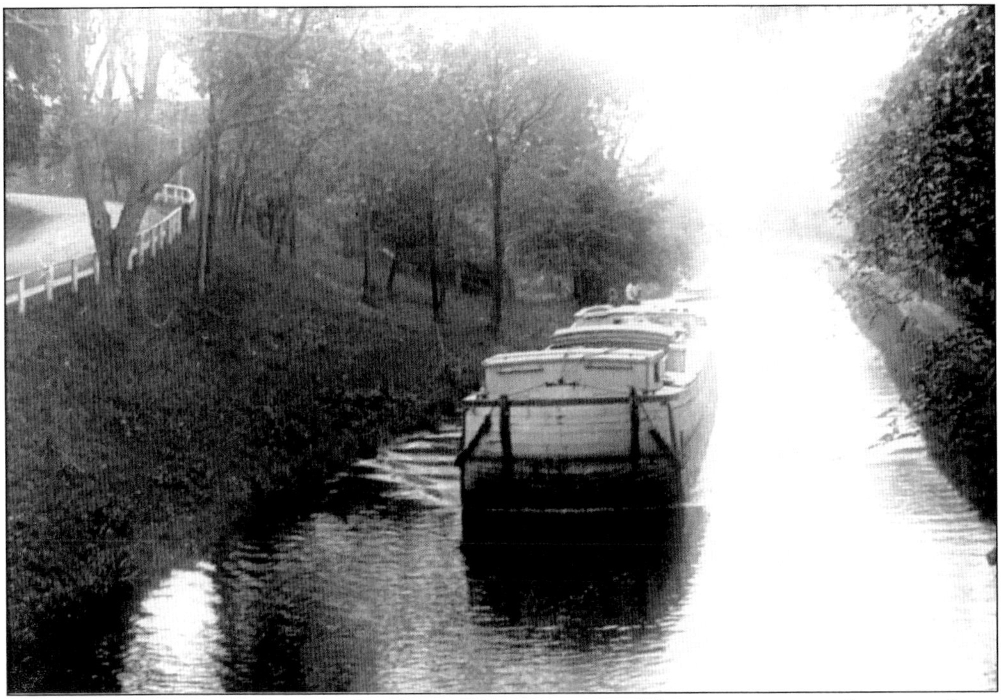

The *Dorris E. Lane*, a Port Leyden boat, advances toward Rome at a rate of two to three miles per hour. The location is Robbers' Hill outside Westernville. (Courtesy of New York State Museum.)

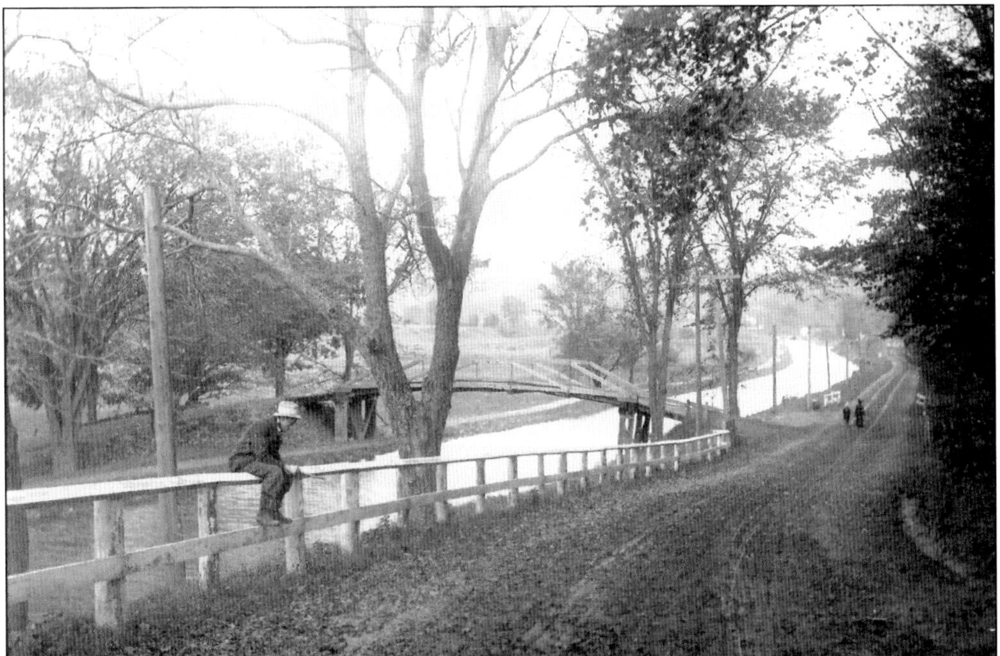

The scene is tranquil, but legends tell of bandits at Robbers' Hill who preyed on passing canal boats. Perched on the fence is George Spinks Jr. (Courtesy of Mary Centro.)

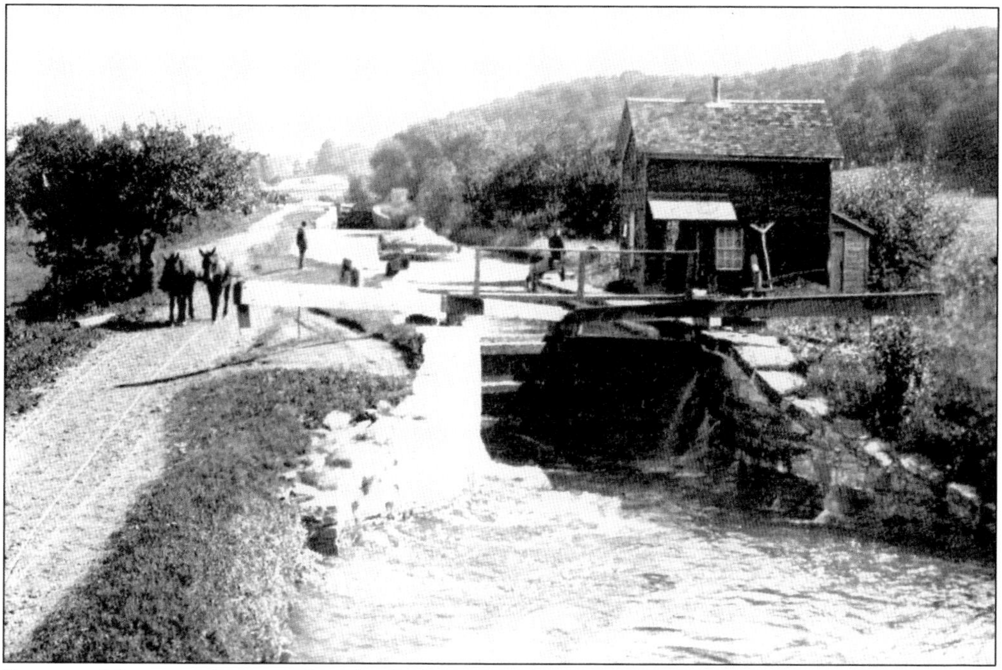

Boatmen blew a tin horn to alert the lock tender of their approach. The typical lock had a lift of 10 feet, and a boat could be locked through in about 15 to 20 minutes. Originally, lock tenders were on duty 24 hours a day; in later years, workdays were shortened to 12 hours.

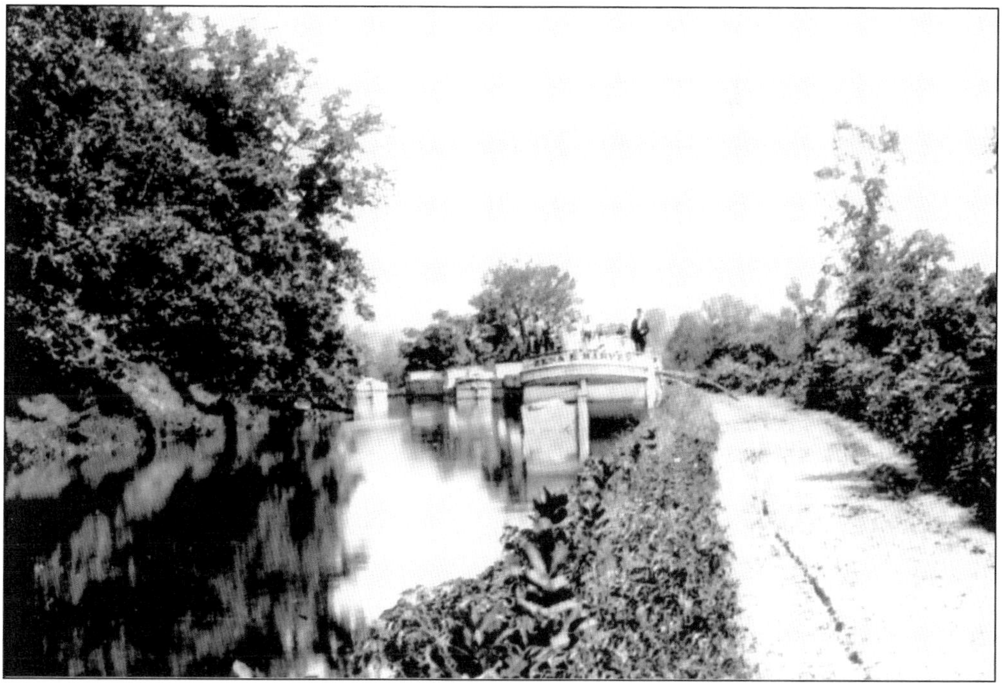

Boats queue up at a lock. The last boat in line is the *Zena E. Harvey*, named for the granddaughter of its captain, Henry Harvey. Lewis Seiter of Boonville built this boat at Theodore B. Basselin's boatyard in Castorland.

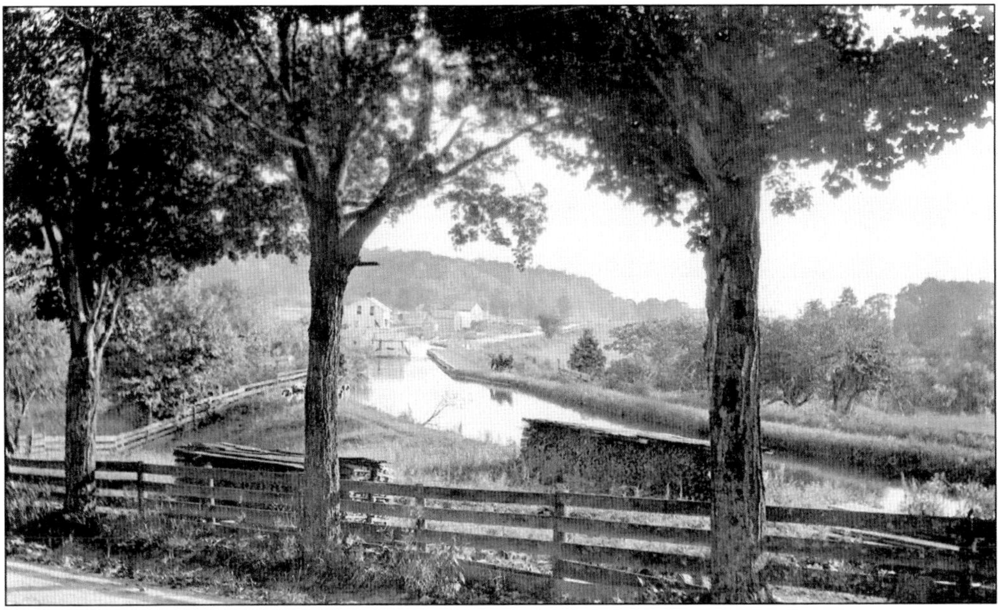

In the distance, a canal boat crosses the aqueduct at Frenchville, about 10.5 miles north of the Erie Canal. (Courtesy of Mary Centro.)

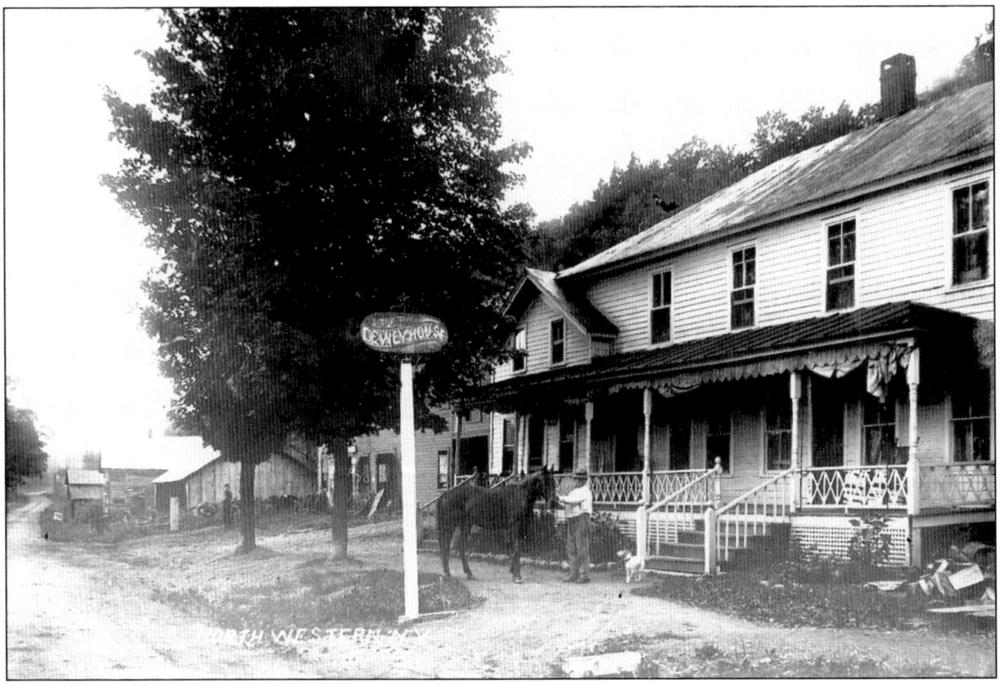

North Western was a thriving canal town about halfway between Rome and Boonville. Charles Dorrity's Dewey House is typical of the many hotels that appeared along the canal. For a time, Dorrity served as superintendent of Section 1, the southern portion of the canal.

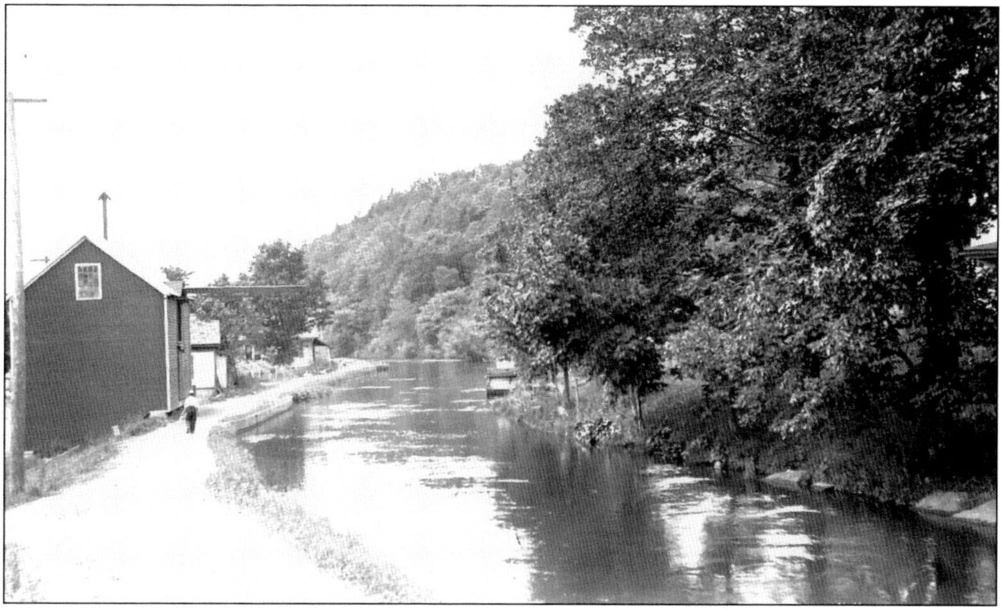

Jerome V. Gue of North Western was one of the earliest boatbuilders on the Black River Canal. His dry dock and boatyard were on the southern edge of the village. The building on the left is a storehouse for feed. (Todd Weseloh Collection.)

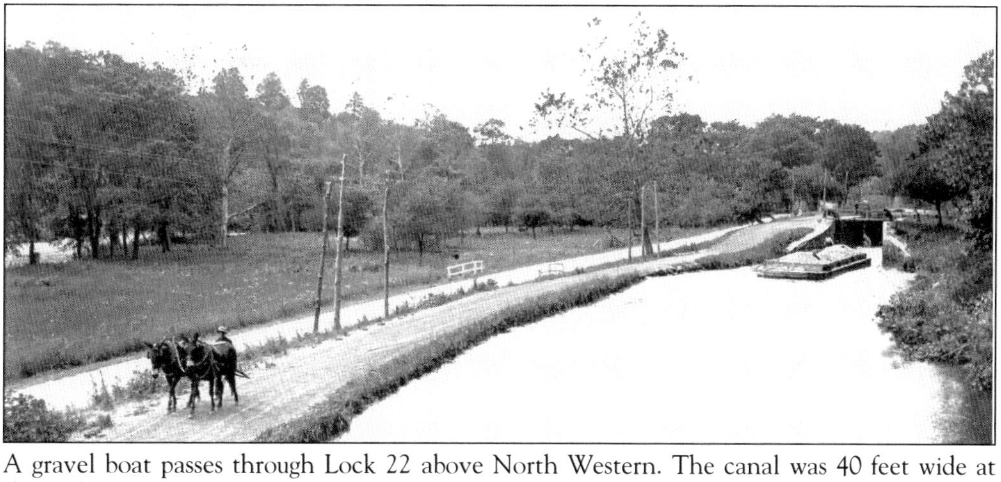

A gravel boat passes through Lock 22 above North Western. The canal was 40 feet wide at the surface and 28 feet wide at the bottom. Water depth was supposedly 4 feet but varied with the weather. In drought years, canal operations might be suspended for weeks on end. (Todd Weseloh Collection.)

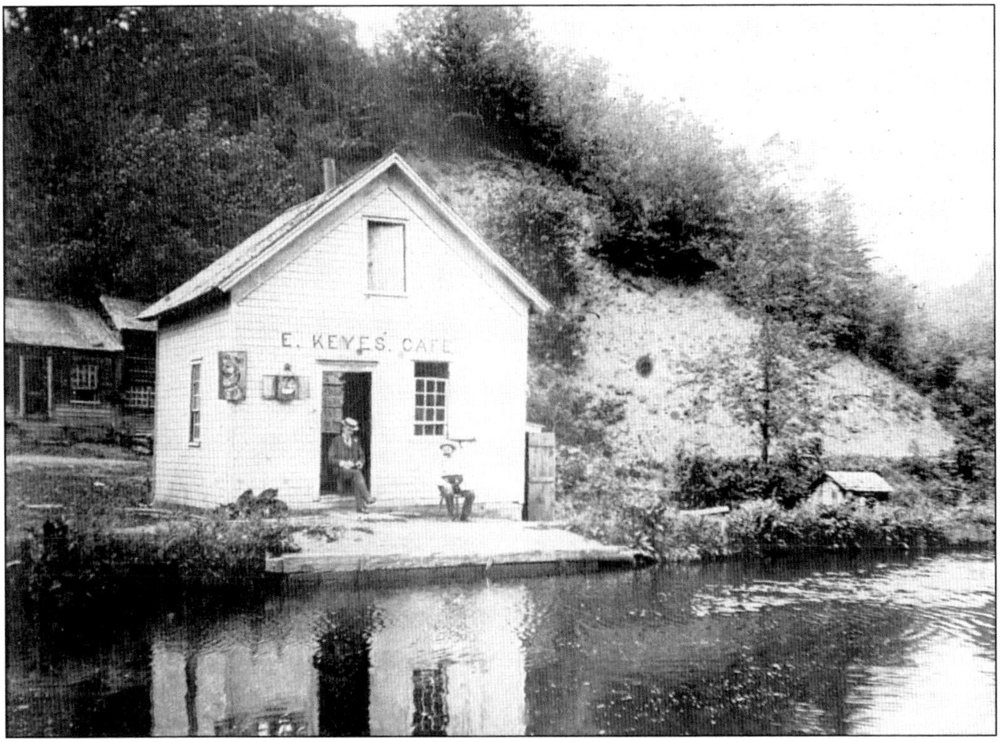


At Dunn Brook (Lock 31), Ed Keyes (sitting right) operated a well-known canal-side café and grocery store. Keyes owned a canal boat, the *Mina*, before starting his café. (Courtesy of Ken Clark.)

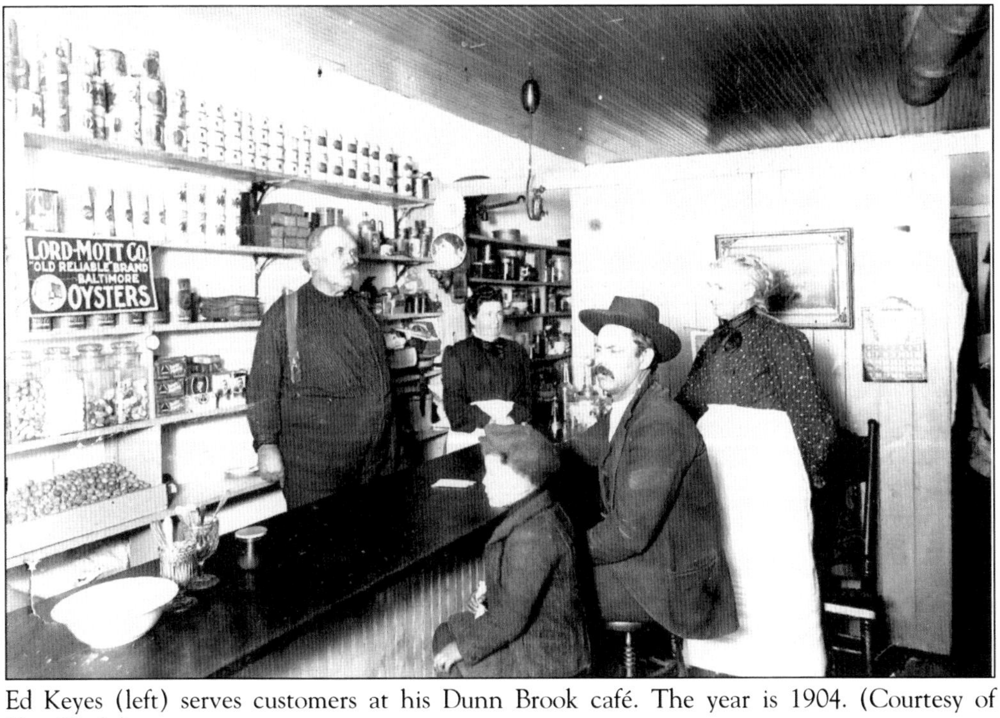


Ed Keyes (left) serves customers at his Dunn Brook café. The year is 1904. (Courtesy of Ken Clark.)

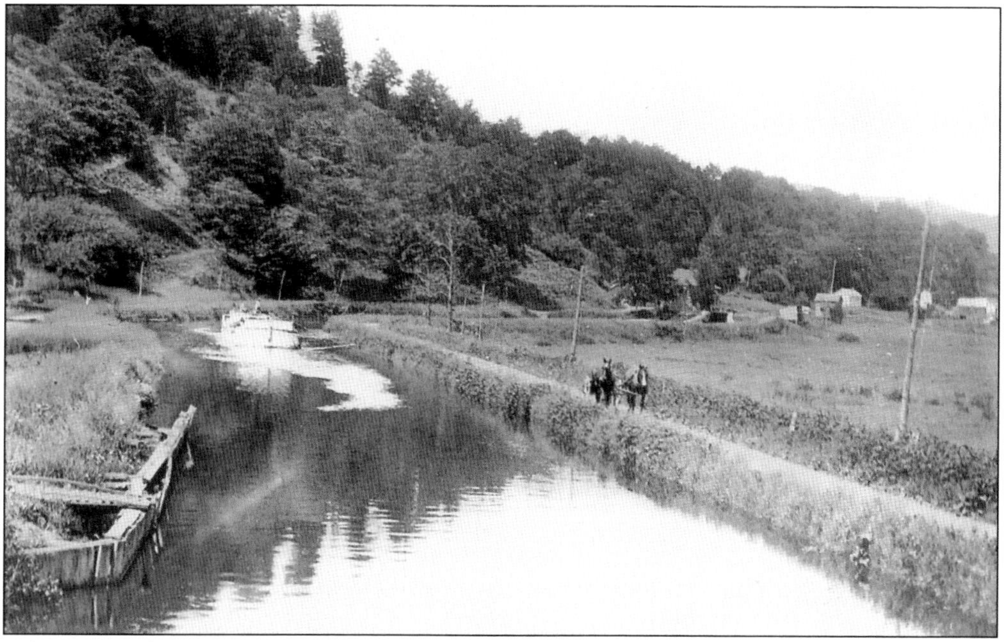

A boat exits the southern end of the Lansing Kill Gorge near Dunn Brook. On average, a lock would be needed every one-fifth of a mile between Dunn Brook and the summit level.

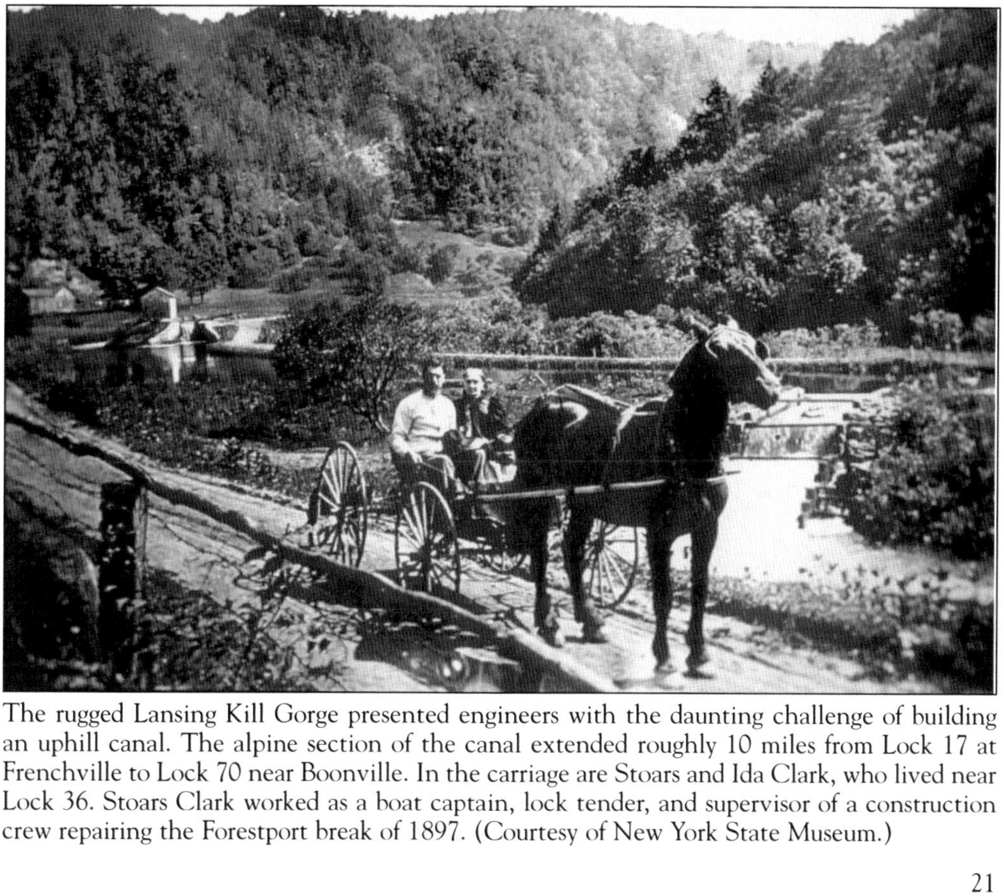

The rugged Lansing Kill Gorge presented engineers with the daunting challenge of building an uphill canal. The alpine section of the canal extended roughly 10 miles from Lock 17 at Frenchville to Lock 70 near Boonville. In the carriage are Stoars and Ida Clark, who lived near Lock 36. Stoars Clark worked as a boat captain, lock tender, and supervisor of a construction crew repairing the Forestport break of 1897. (Courtesy of New York State Museum.)

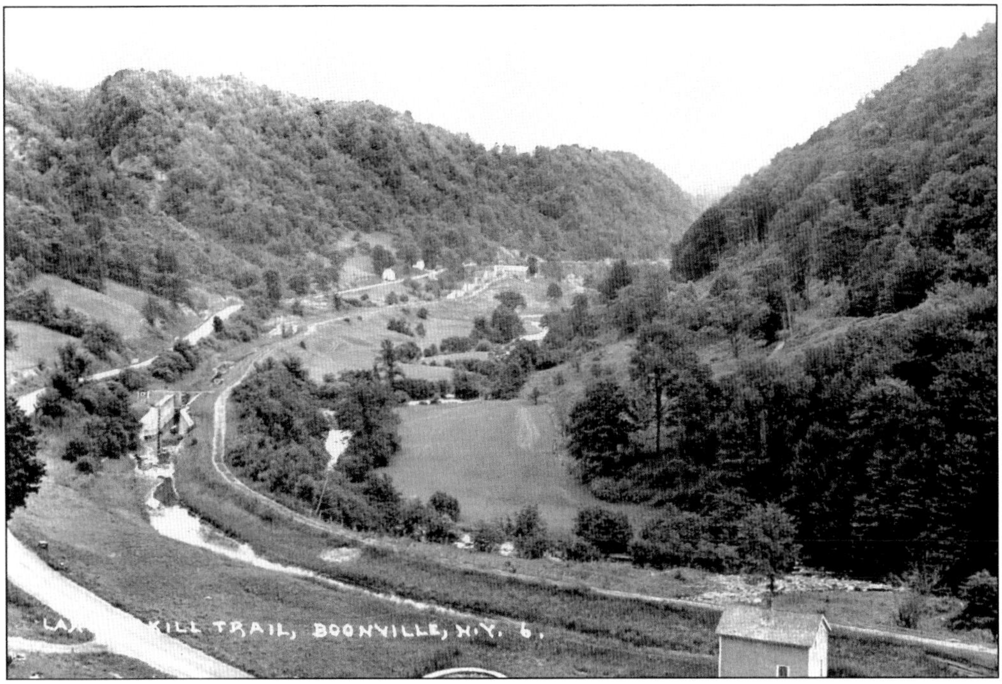


Formed by glacial runoff during the last ice age, the Lansing Kill Gorge widens near Locks 37 and 38. Before the canal could be excavated, the forest had to be cleared and stumps had to be uprooted. The hills above the canal rise to more than 1,500 feet.

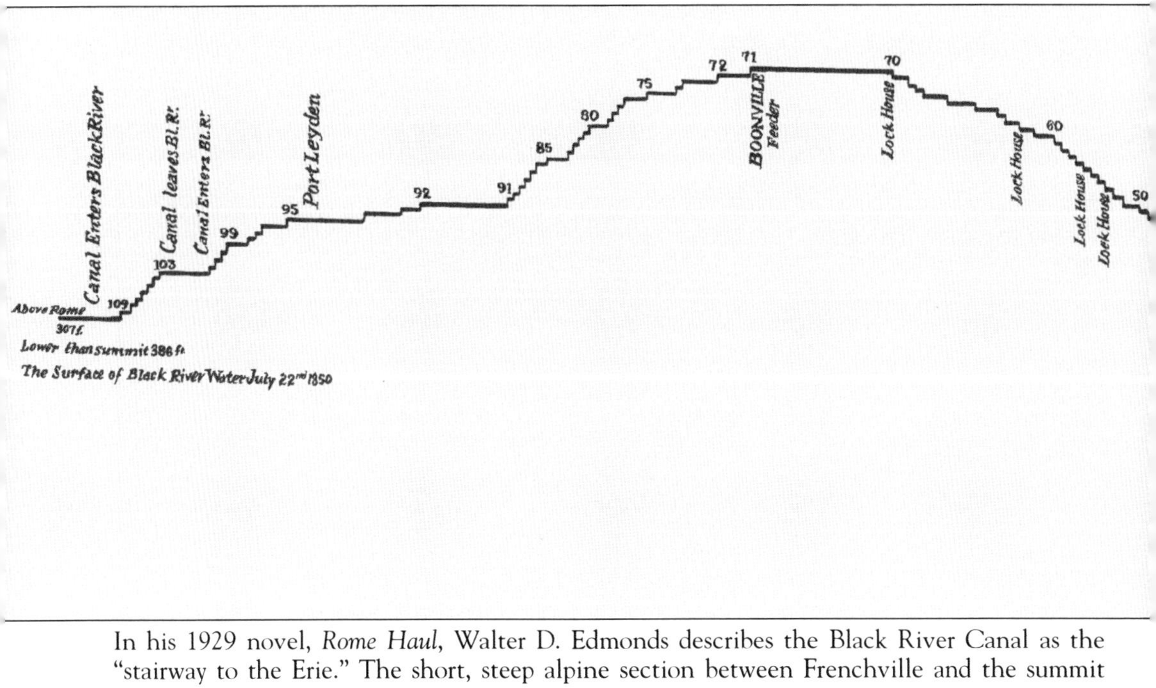


In his 1929 novel, *Rome Haul*, Walter D. Edmonds describes the Black River Canal as the "stairway to the Erie." The short, steep alpine section between Frenchville and the summit

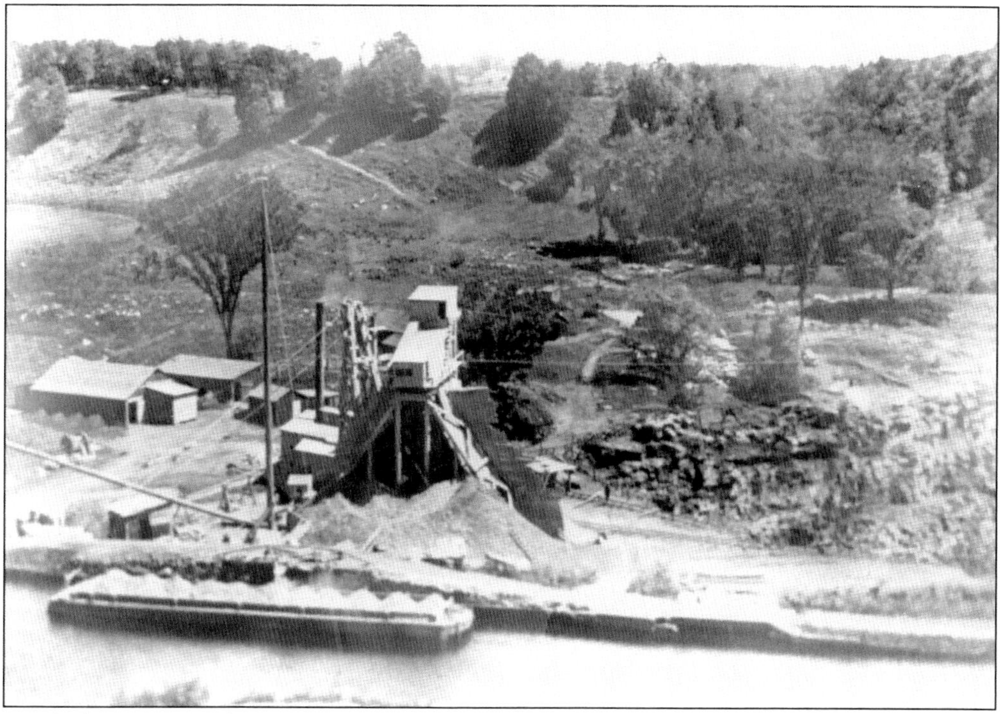

Crushed stone is loading at Hough's quarry in the Lansing Kill Gorge. Here, Utica shale sits atop Trenton limestone. This limestone proved adequate for the locks and was quarried at or near each lock site. Sinkholes, however, frequently formed in areas where the canal bed was carved out of shale.

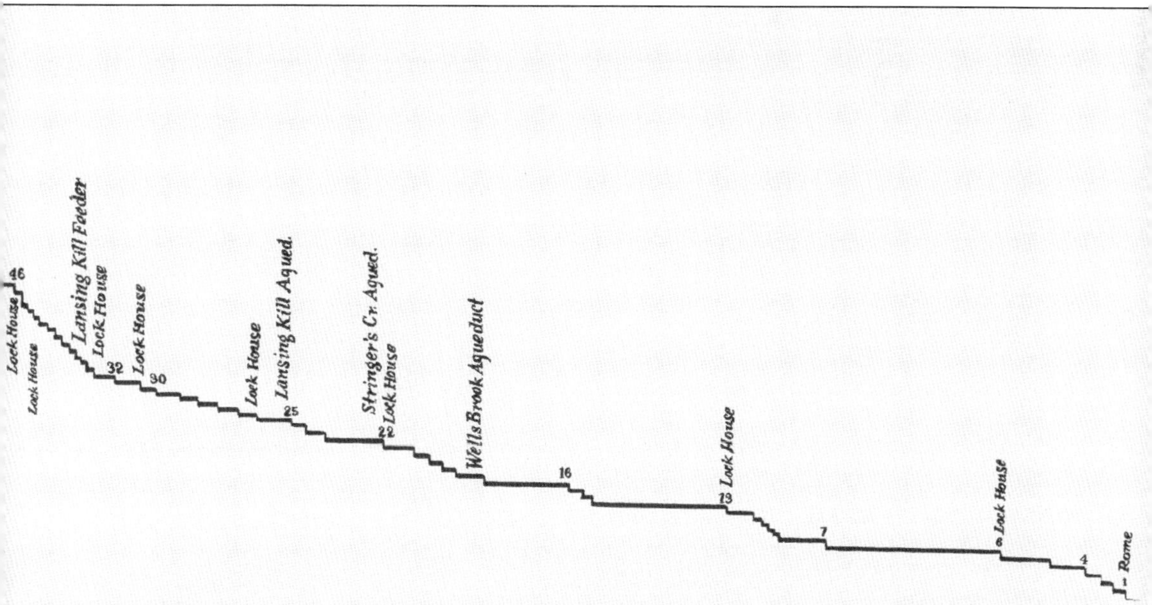


level required 53 locks. The diagram is from the New York State Canal Commissioners' *Annual Report* for 1859.

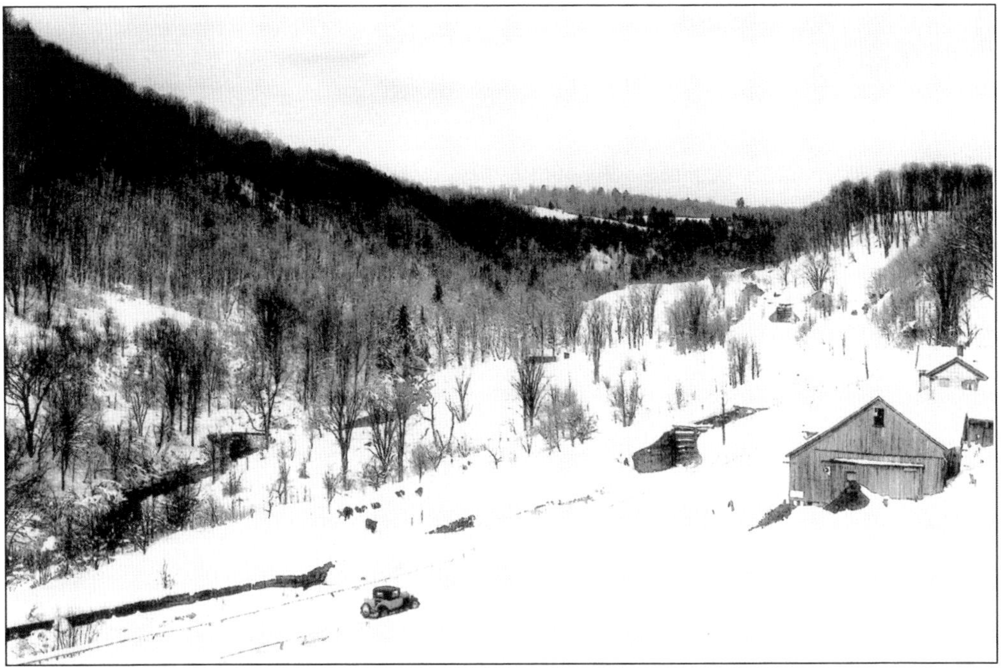

Melting snow and heavy spring rains in the Lansing Kill Gorge regularly flooded the canal with silt and rubble, damming the water and causing the earthen walls to break. The photograph dates from the late 1920s.

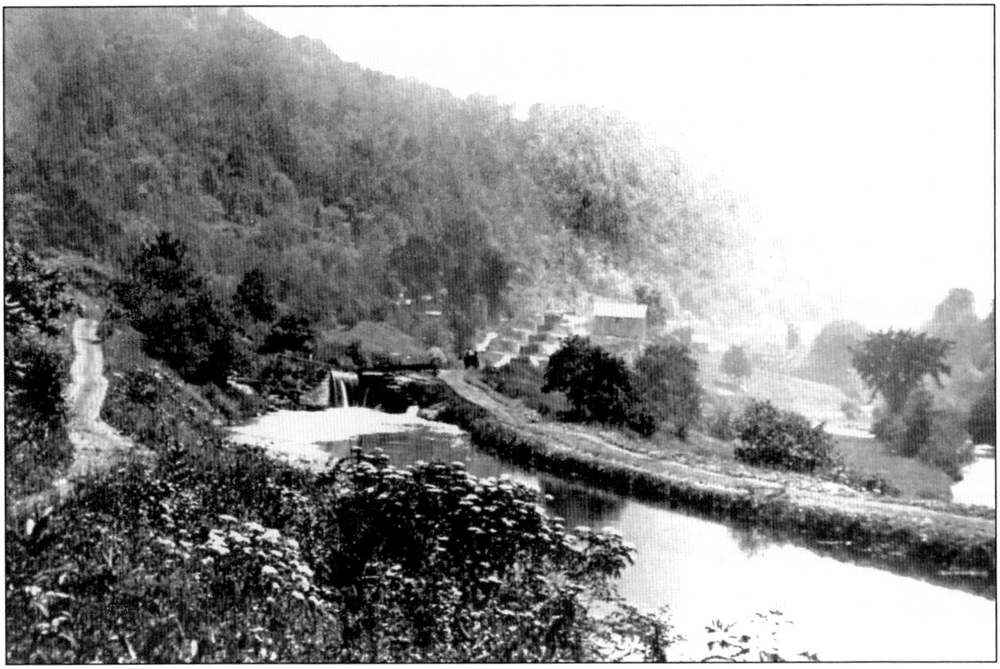

The Five Combines was a celebrated achievement in 1850. Located 7.5 miles south of Boonville, this flight of five adjoining locks (39–43) was capable of lifting boats 50 feet. Between 1853 and 1857, the Ogdensburg, Clayton, and Rome Railroad company attempted, but failed, to build a railroad through this difficult terrain.

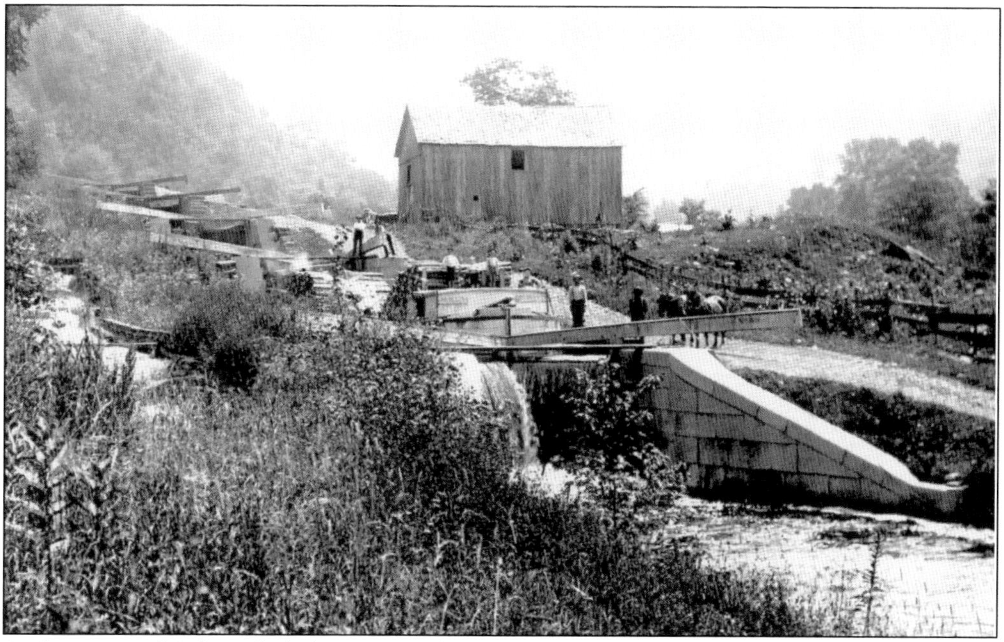

A shipment of Adirondack logs passes down through the Five Combines on its way south. Closed gates are angled upstream for strength and pressed together by stream flow. Valves at the bottom of the gates allowed the lock tender to fill or empty the lock. The lock tender here is Roscoe Clark (far right), and his assistant is Albert Remp (far left).

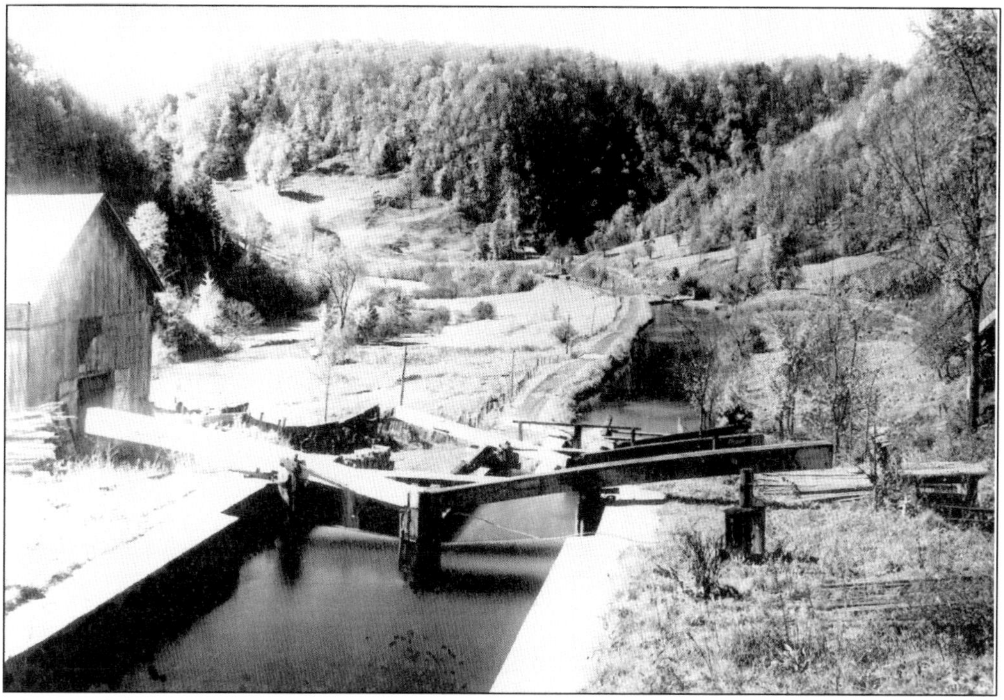

The canal flows 100 feet above the Lansing Kill at the Five Combines. Lock gates open and close by means of balance beams, large tapered timbers resting on a pivot and attached to the gate. Mitre sills, heavy timbers on the bottom of the lock, brace the closed gates.

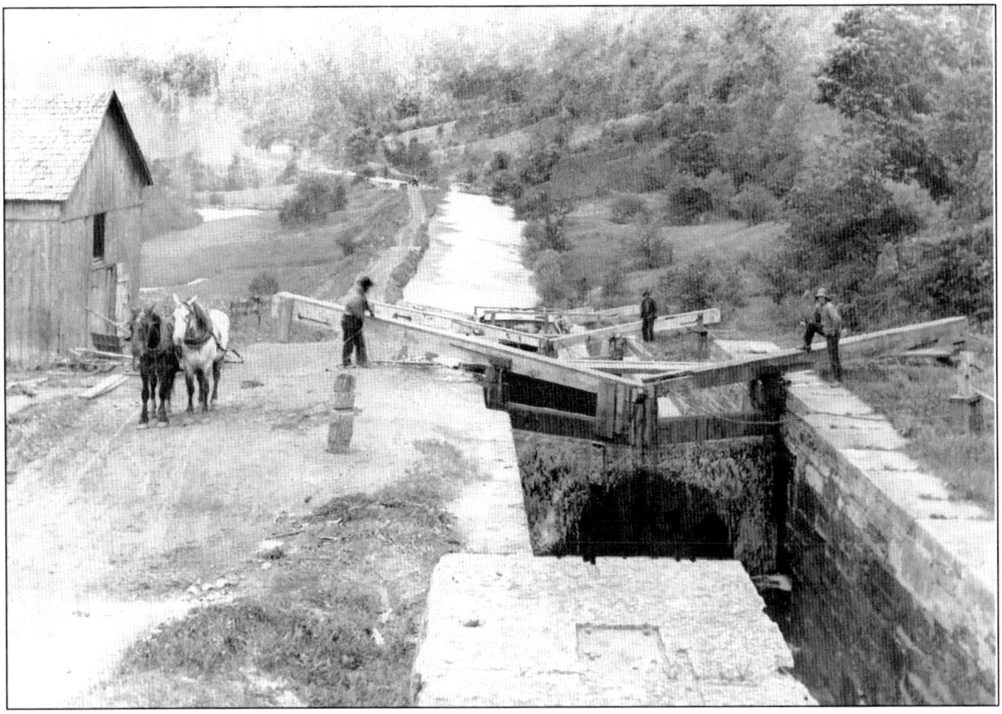

Recesses were built into the locks so that, once opened, gates would rest evenly with the lock walls. The two gates did not work in tandem. After opening one gate, the lock tender crossed the lock on a plank and opened the gate on the other side. (From Rome Historical Society's Photographic Collection.)

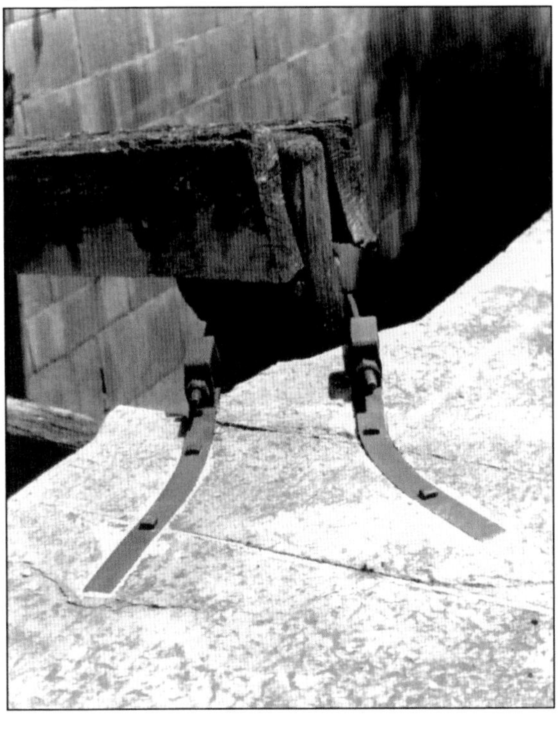

Each gate was attached to a large quoin post. A metal strap at the top held the quoin tightly against the stone wall. A metal plate with a protruding iron rod was attached at the bottom to form a pivot. The rod was anchored in wooden timbers bolted into the stone foundation.

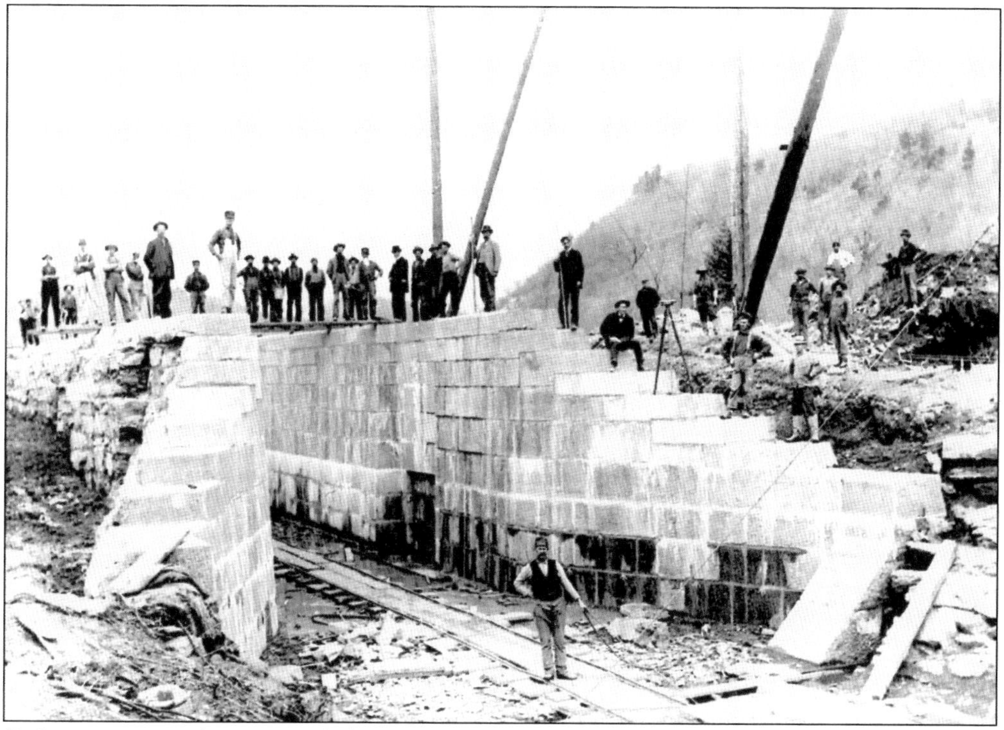

Ordinary wear and tear, coupled with the harsh climate, meant that locks were in need of frequent repair. The stone walls of the lock are buttressed in three places, and they are thicker on the ends.

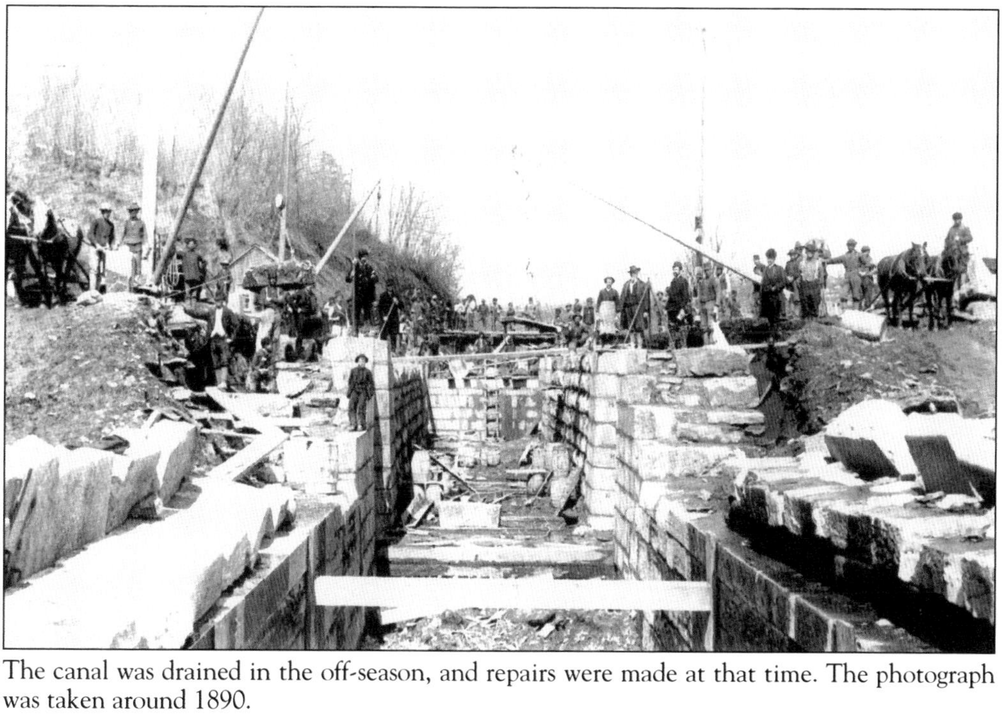

The canal was drained in the off-season, and repairs were made at that time. The photograph was taken around 1890.

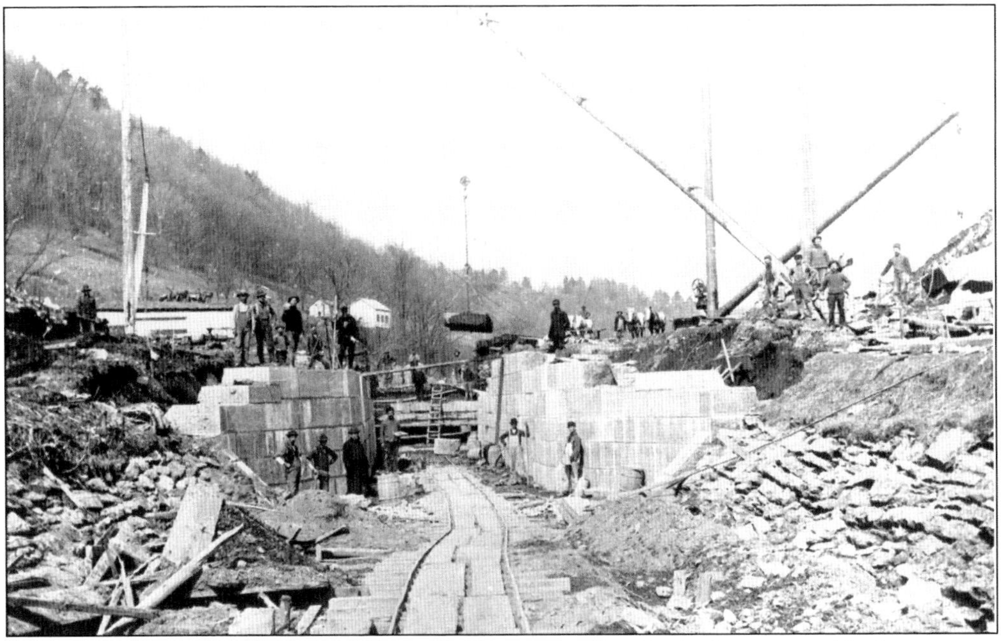

From time to time, locks had to be partially or entirely rebuilt. Limestone blocks sat on a foundation of timbers which, in turn, rested on wooden pilings driven into the ground. The Five Combines was largely complete at the time of the Stop Law in 1842. Portions of the Five Combines were rebuilt in 1881, 1895, and 1912.

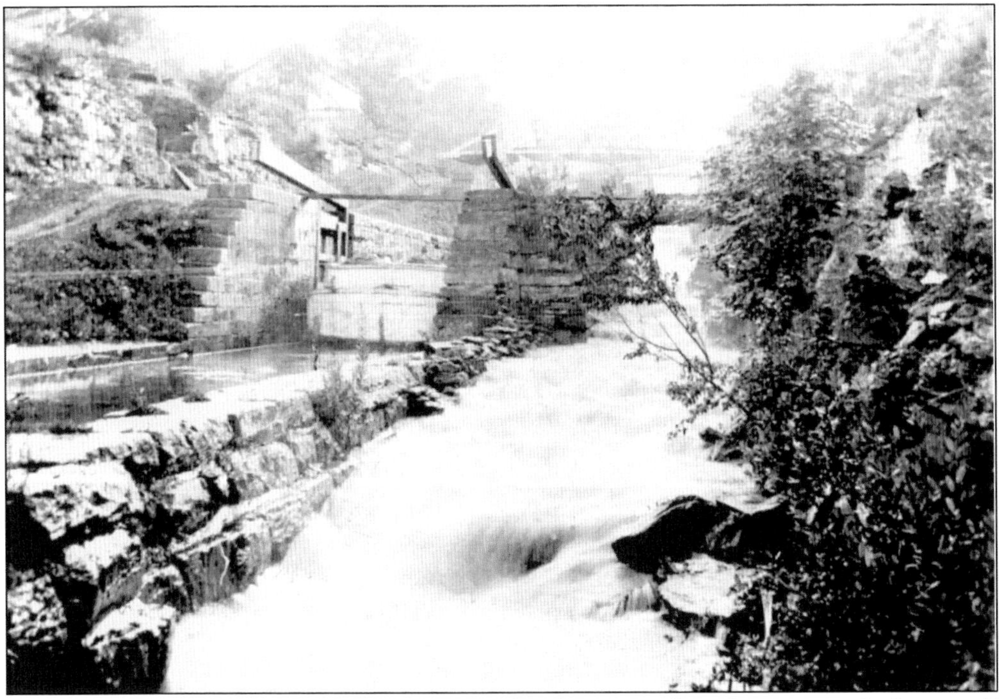

In places, the Lansing Kill Gorge was so narrow that the stream flowed up against the berm wall of the canal, as it does here at the lower three combines (Locks 44–46). Another 25 locks were located between here and the summit at Boonville.

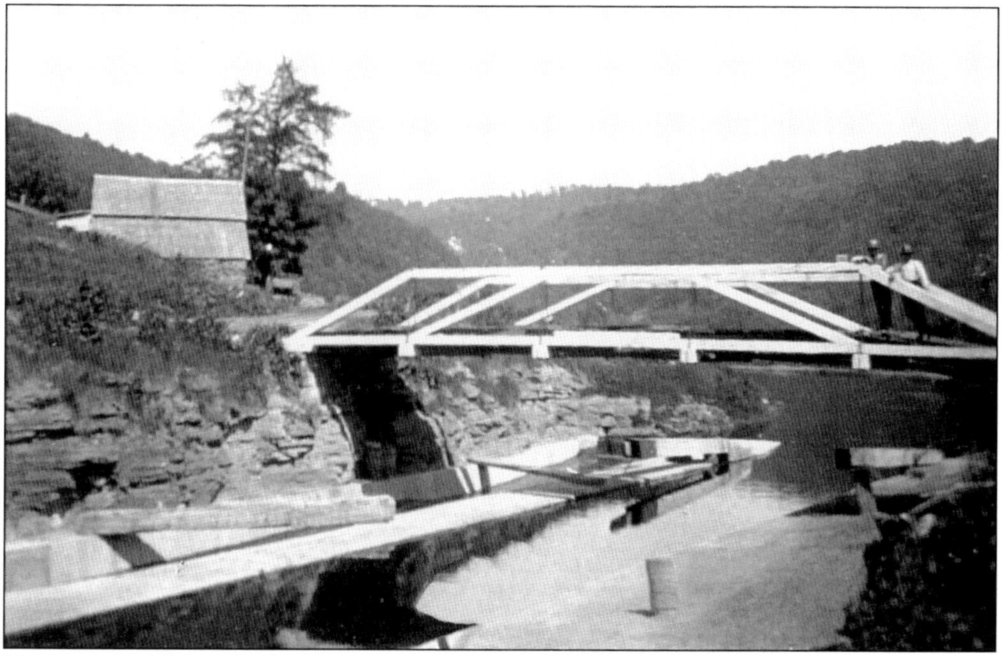

This trapezoidal truss bridge was one of 66 bridges spanning the overland section of the Black River Canal in 1892. The bridge sits above the lower three combines, about 18 miles north of the Erie Canal. Marty Clark was lock tender when this photograph was taken. (Courtesy of New York State Museum.)

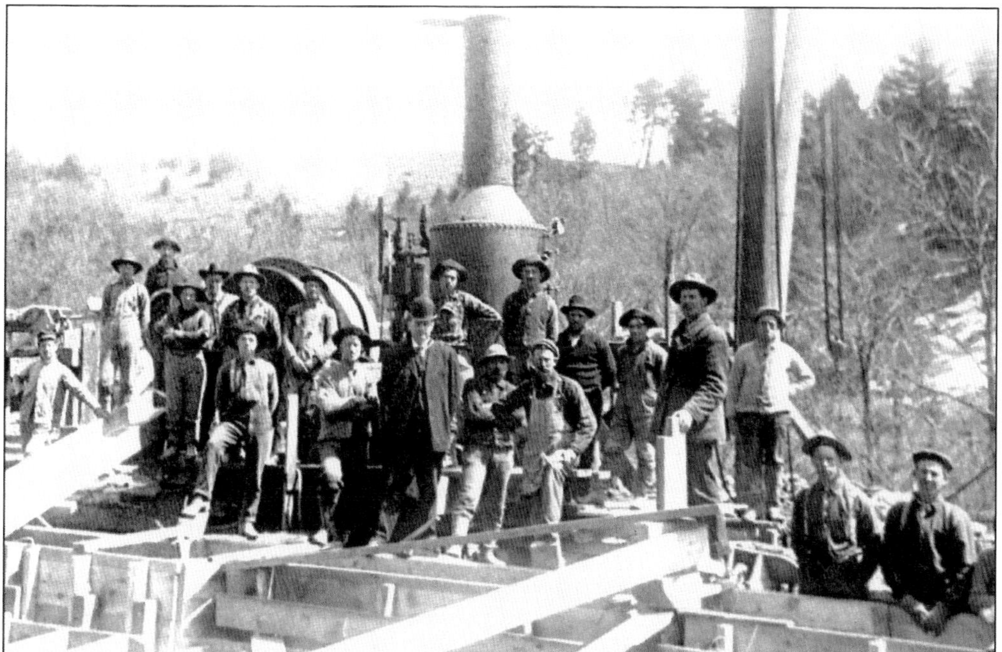

With the help of steam power and a smartly dressed supervisor, a work crew rebuilds a lock in the Lansing Kill Gorge. (Courtesy of New York State Museum.)

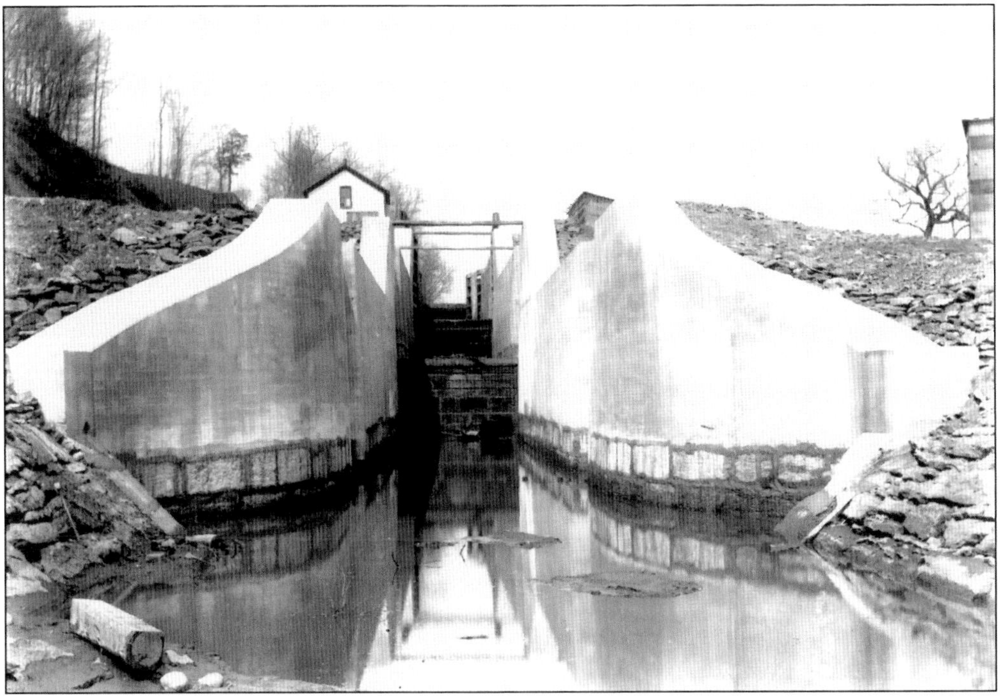

Limestone blocks were joined and sealed with hydraulic cement patented in 1820 by Canvass White of Whitestown. In later years, concrete was used to replace limestone blocks, as it is here at the upper three combines (Locks 58–60).

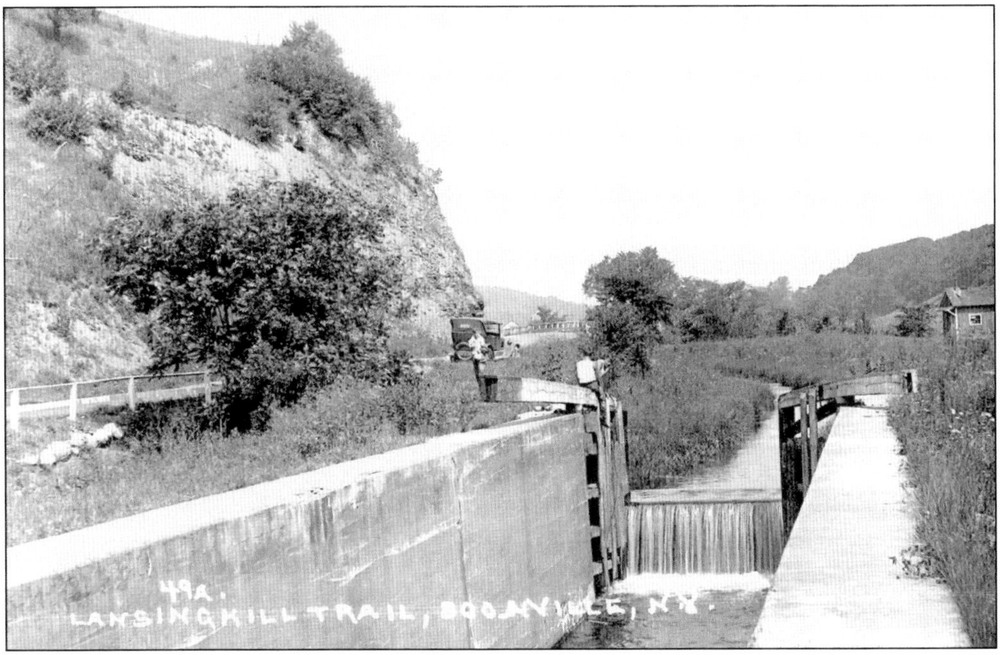


Three young men stop at a resurfaced lock in the upper Lansing Kill Gorge. In November 1870, George Crofoot hauled a load of cheese through this lock on his way from Lyons Falls to Rome. The 35-mile trip took 27 hours, a record at the time. (Courtesy of Vincent T. Fitch.)

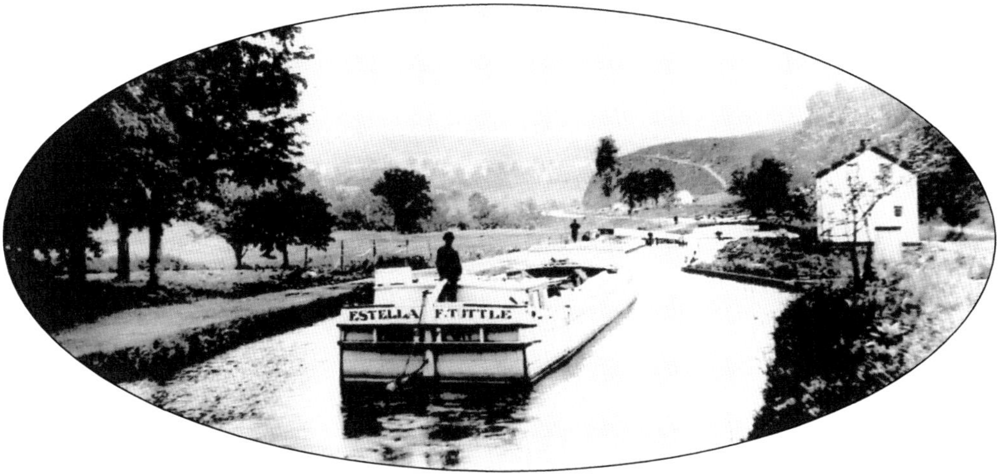


The *Estella Tuttle* approaches Lock 70 at the northern end of the Lansing Kill Gorge—with distant vistas only hinting at the lockage ahead. The *Estella Tuttle* was captained by Warren Tuttle and owned by the Van Amber brothers of Beaver River.

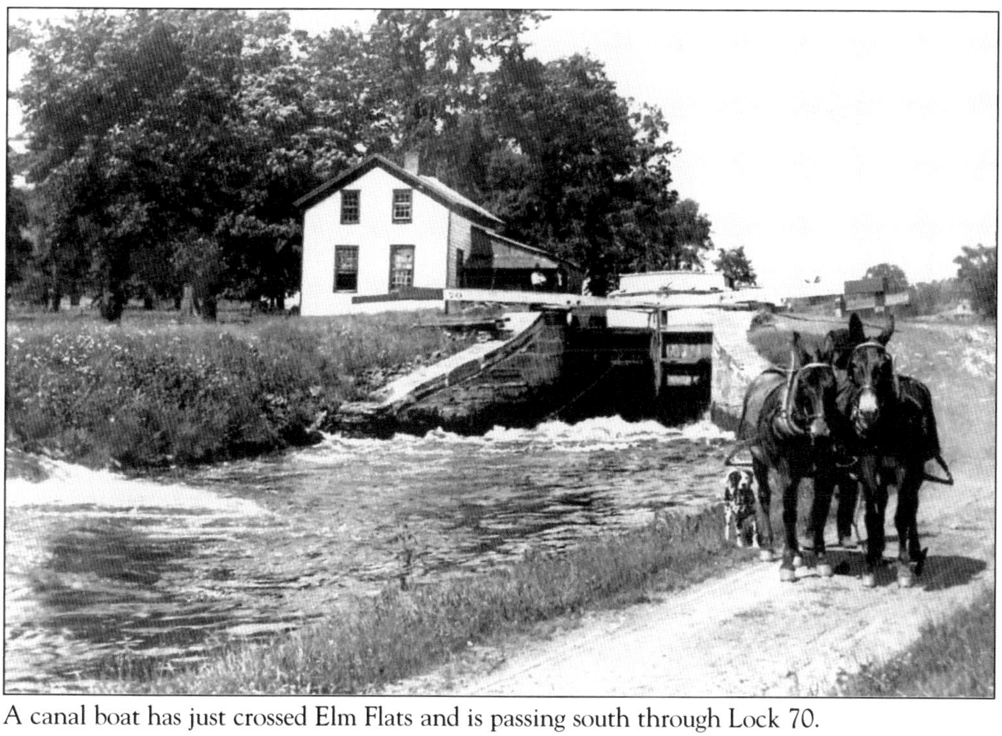

A canal boat has just crossed Elm Flats and is passing south through Lock 70.

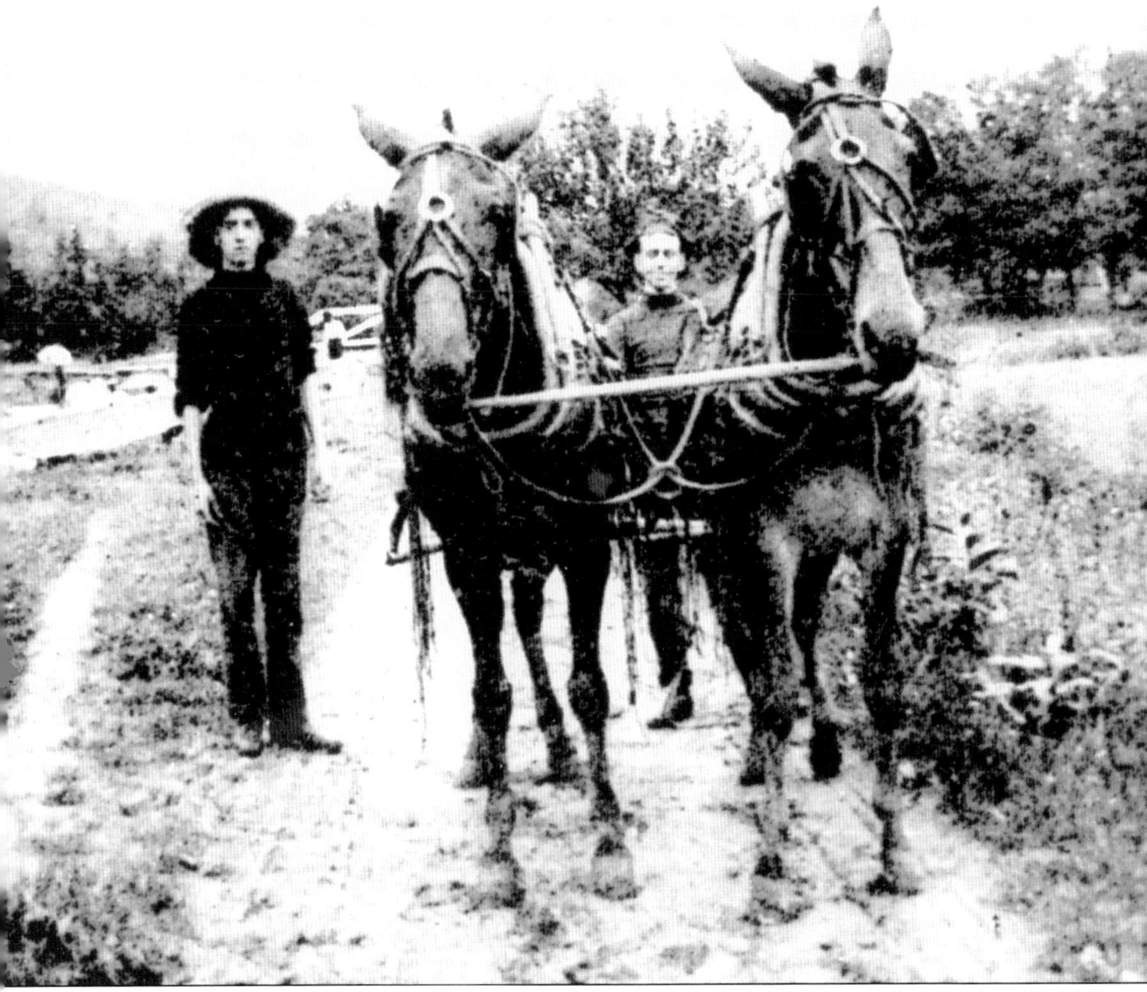

Horses and mules were both used on the Black River Canal. Many canallers preferred mules because they were thought to be smarter and sturdier. Moreover, mules, unlike horses, could eat from a feedbag while working, and they refused to drink contaminated water.

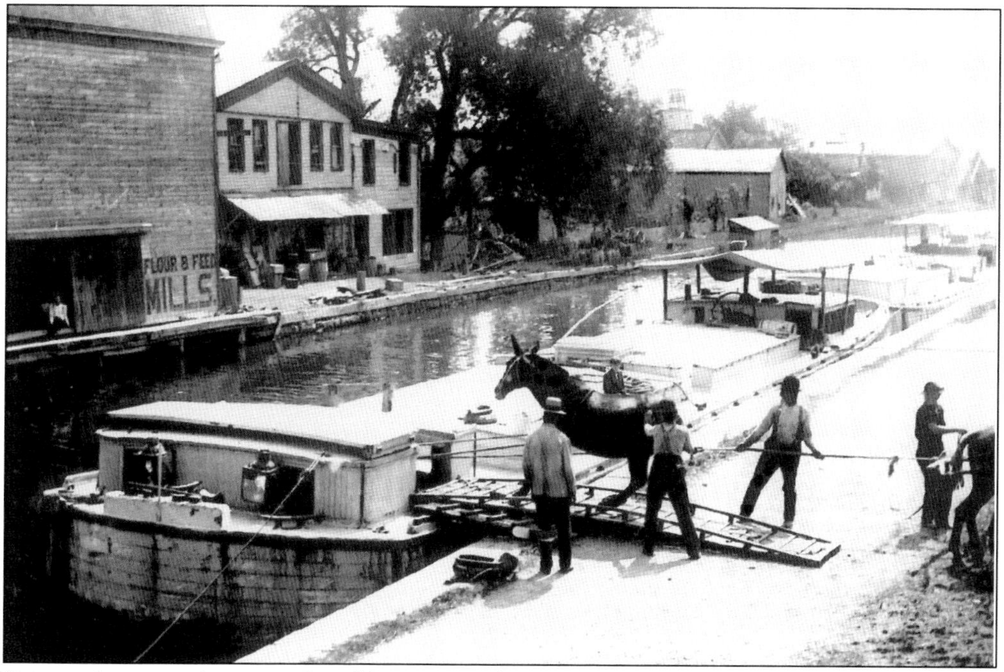


Drivers change mule teams. Canal boats carried two teams of mules, with one team working six hours as the other team rested. After crossing an oaken horse bridge, the mule walked down a cleated ramp into a 9-by-12-foot bow stable. This scene is at Durhamville on the Erie Canal.

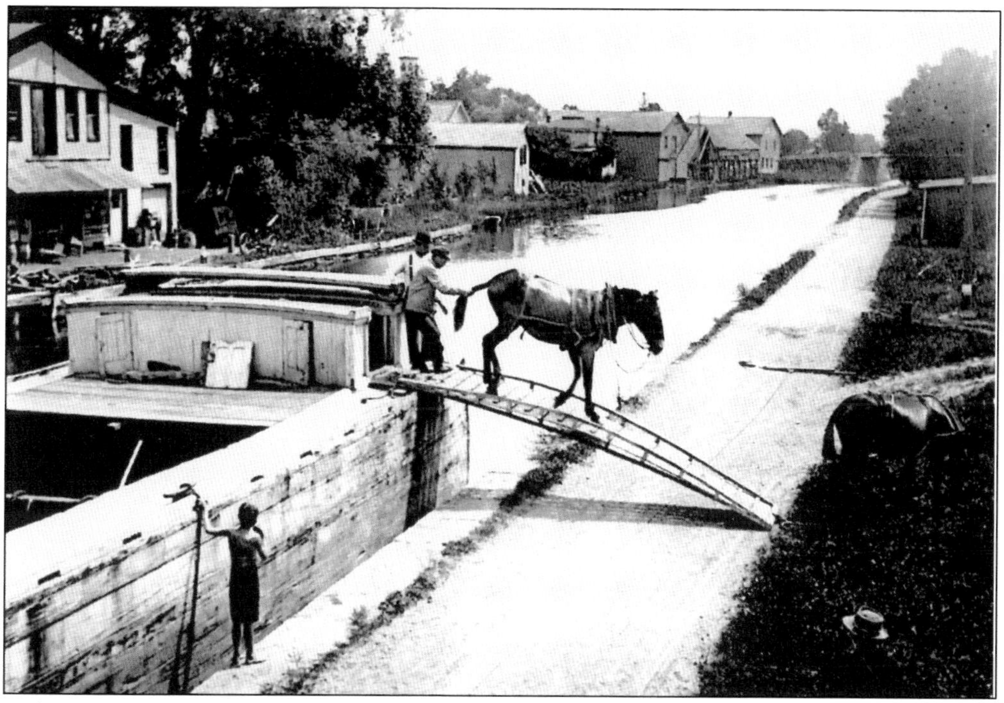

A driver "tails off" a mule to steady it and keep it from going too fast. The workday began before dawn, when a youngster or hired hand rose to feed, groom, and harness the mules. And the day could be long, lasting until 9:00 or 10:00 p.m.

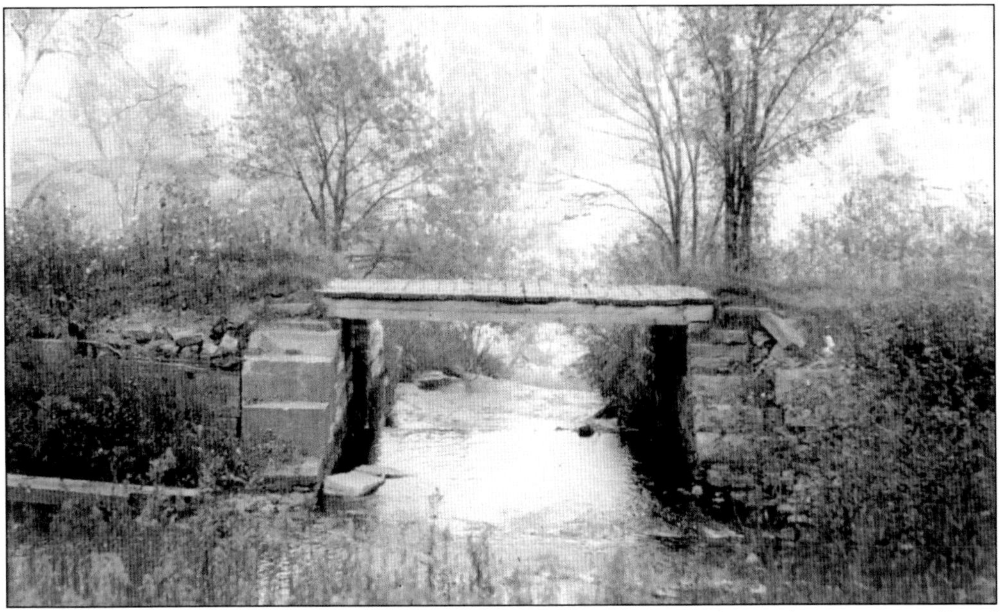

Just south of Lock 70, a waste weir fed surplus water from the canal into the Lansing Kill and thence into the Mohawk River and Erie Canal. A dam on the Lansing Kill near Lock 34, however, allowed water to be diverted back into the canal as needed. Today, Lock 70 is as far south as water flows through the canal. The photograph dates from the 1960s. (Courtesy of New York State Museum.)

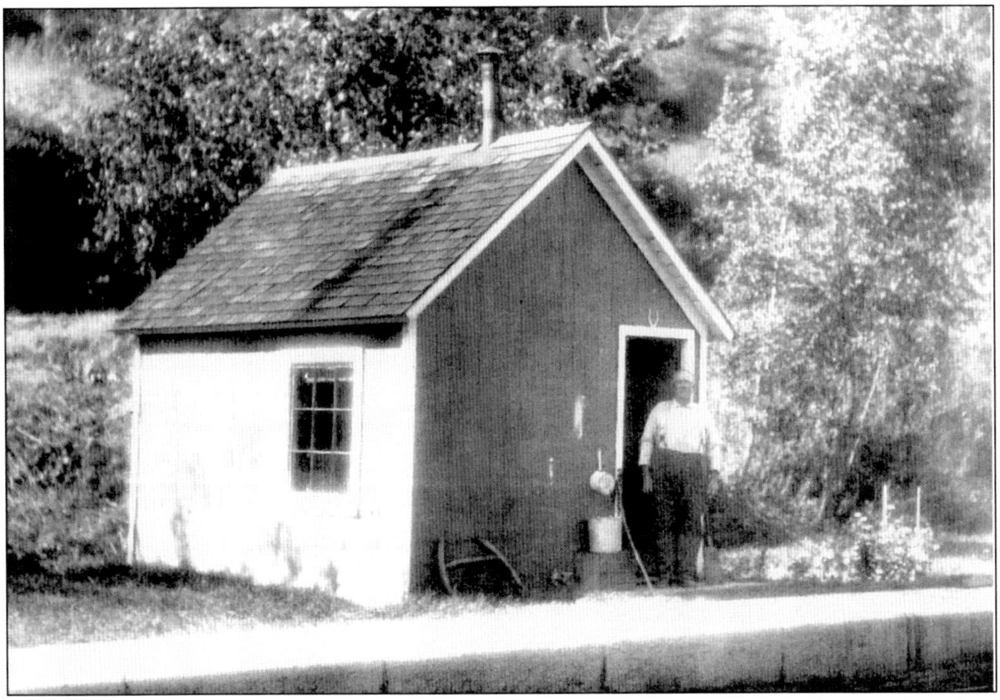

The state supplied housing at each lock or combine. Some structures were shanties furnished simply with a table, chairs, and built-in bunk bed. Lock tenders depended on political patronage for their jobs.

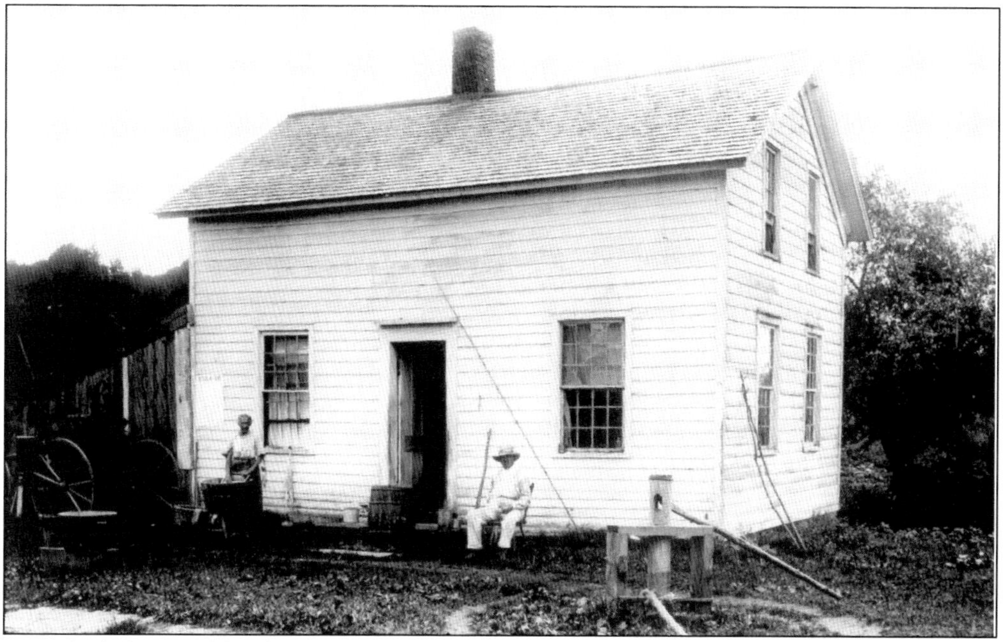

Some lock houses, like this one at Lock 13, could accommodate a small family. Here, a lock tender's wife washes clothes as her husband sits in front of their two-story home. Regulations are posted on the left of the building. The capstan on the right was used to pull boats into the lock chamber.

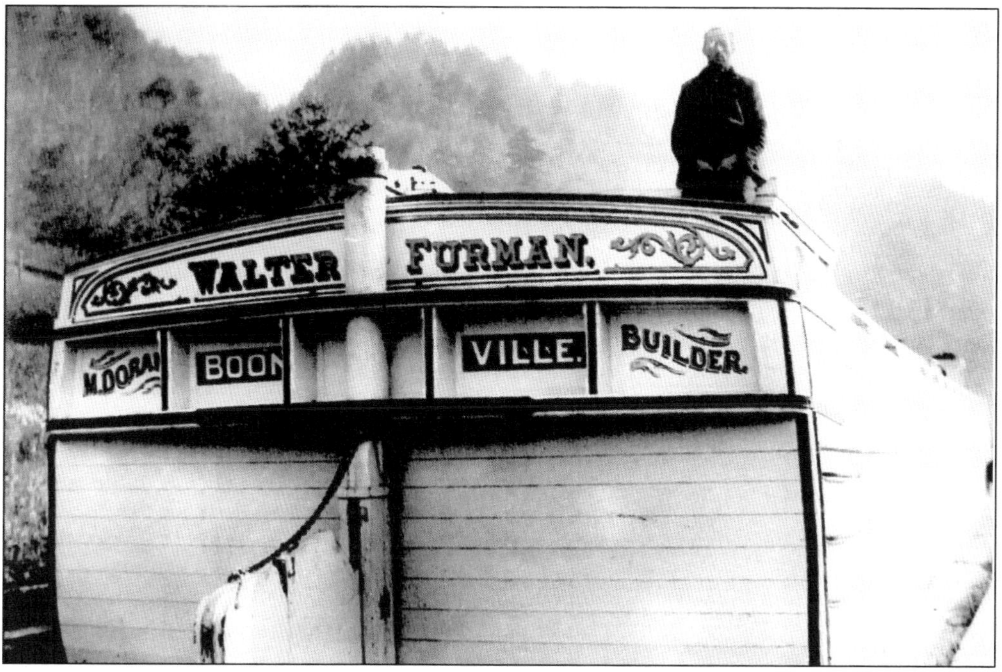


Walter Furman shipped potatoes that he grew on his Black River farm. Michael Doran of Durhamville built the *Walter Furman*. Born in Canada to an Irish immigrant family, Doran had also lived in Boonville and had helped construct the Black River Canal. By the 1890s, he employed 35 men at his Durhamville dry docks, turning out four to six boats a year.

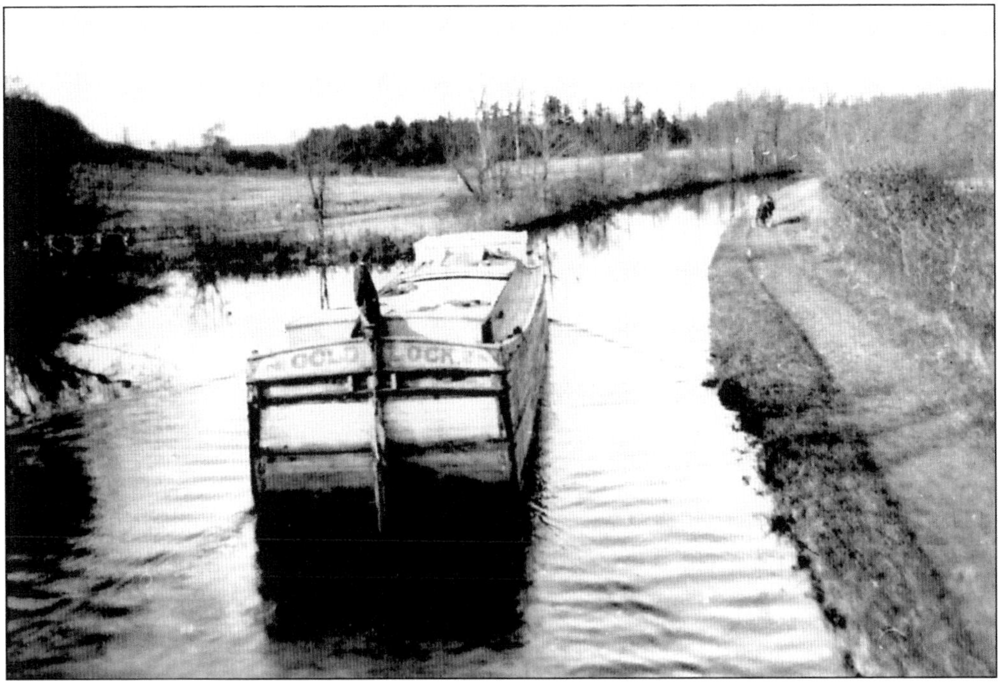

Canal breaks were a constant problem, and a speed limit of four miles per hour was imposed to minimize erosion of the banks. The mules were typically given a 250-foot lead in order to keep the boat from rubbing up against the bank.

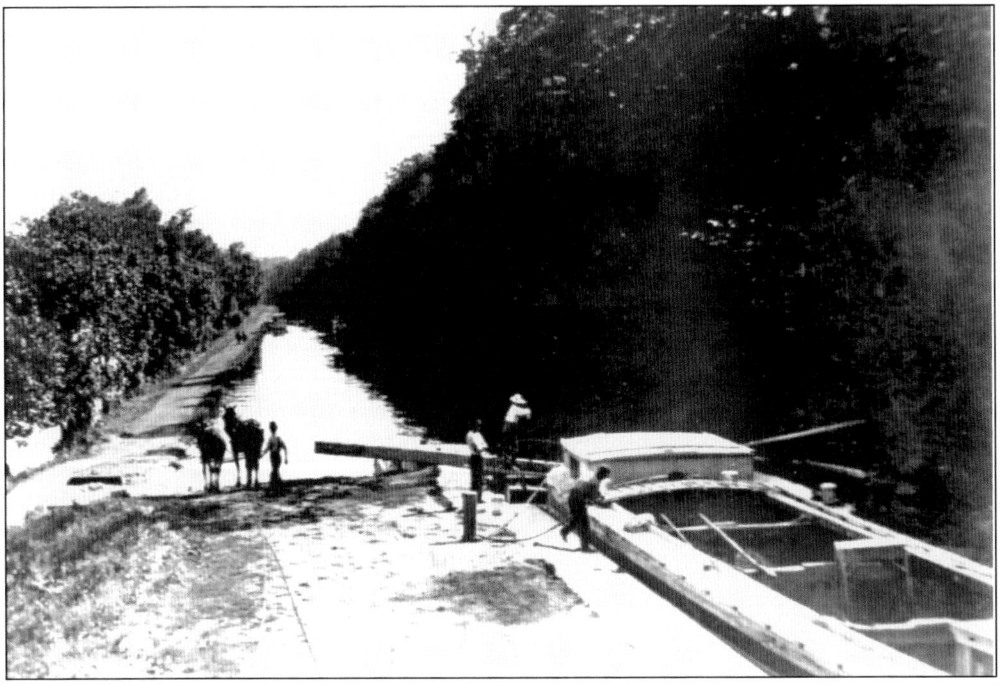

An empty canal boat is locking through near Westernville. Once through, this boat had to yield to the approaching upstream boat. The driver had to stop and slacken his towrope so that the approaching boat could pass over it.

## *Two*

# Forestport and Erie Canal Feeder

Building the canal was one challenge. Finding enough water for it was another. Engineers initially calculated that a 10-mile feeder canal from Forestport to Boonville would deliver enough Black River water to replenish both the Black River Canal and Erie Canal, especially the long level between Utica and Syracuse. From Boonville, two-thirds of the feeder canal's water was to flow south to the summit of the Erie Canal at Rome and one-third was to flow north back into the Black River at Lyons Falls, where it would be available for boat traffic and the mills downriver. Construction of the feeder began in 1838, and the first boat was towed from Boonville to Forestport 10 years later, on December 16, 1848.

With access to the great lumber centers at Albany and the Tonawandas, it was now possible to log the upper Black River Valley profitably, and Forestport soon became an important center of the lumber industry. In 1877, most lateral, or branch, canals were abandoned by the state, but because the Erie Canal was critically dependent on Black River water, the Black River Canal remained open. In 1882, an amendment to the state constitution prohibited the sale or lease of the Black River Canal without a popular vote. From the beginning, however, the water supply at Forestport proved inadequate. To address the problem, the state began damming nearby Adirondack rivers and lakes until a vast impoundment system was formed, capable of holding four billion cubic feet of water. But even this failed to meet the canal system's insatiable needs.

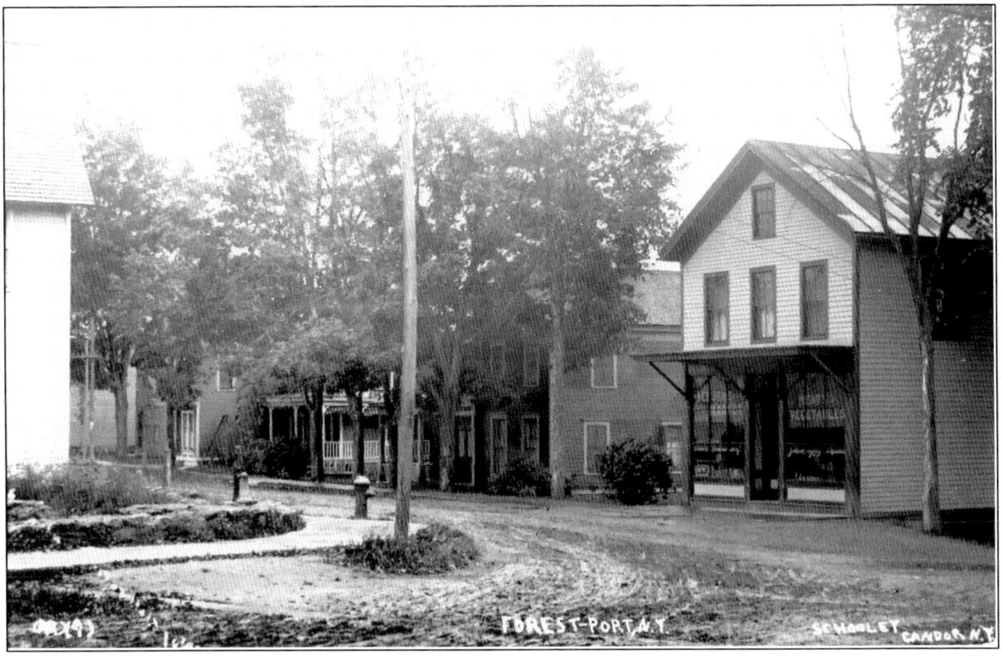


Known variously as Smith's Mills, Punkeytown, and Williamstown, Forestport got the name that stuck when the town became a terminus on the Erie Canal Feeder and logging, in turn, became big business.

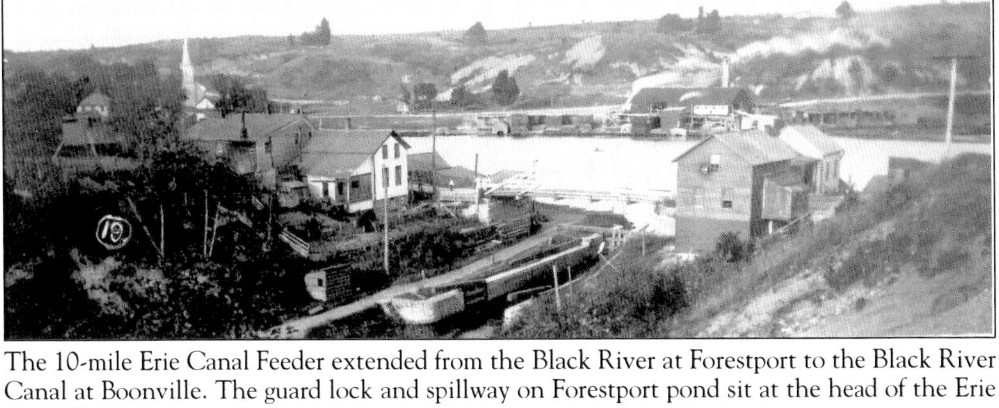

The 10-mile Erie Canal Feeder extended from the Black River at Forestport to the Black River Canal at Boonville. The guard lock and spillway on Forestport pond sit at the head of the Erie Canal Feeder.

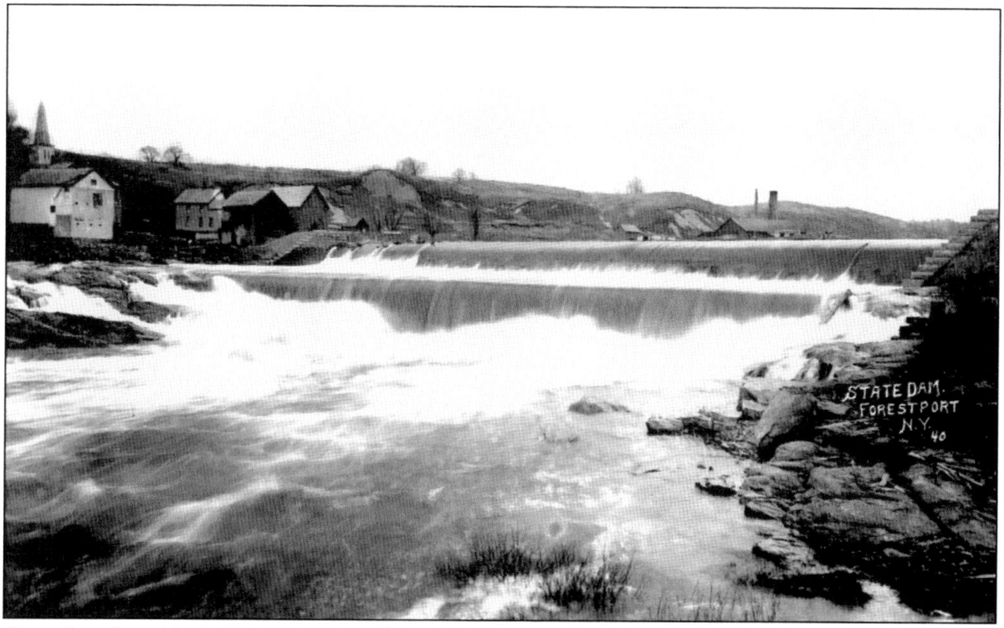


Completed in 1848, the Forestport dam was 18 feet high and 200 feet long. It was designed to raise the Black River 12 feet and provide 16,000 cubic feet of water per minute for 100 days. The dam washed out in the flood of 1869 and was then rebuilt. In 1903, the second wooden dam was replaced with a concrete dam, shown here. (Courtesy of George Davis.)

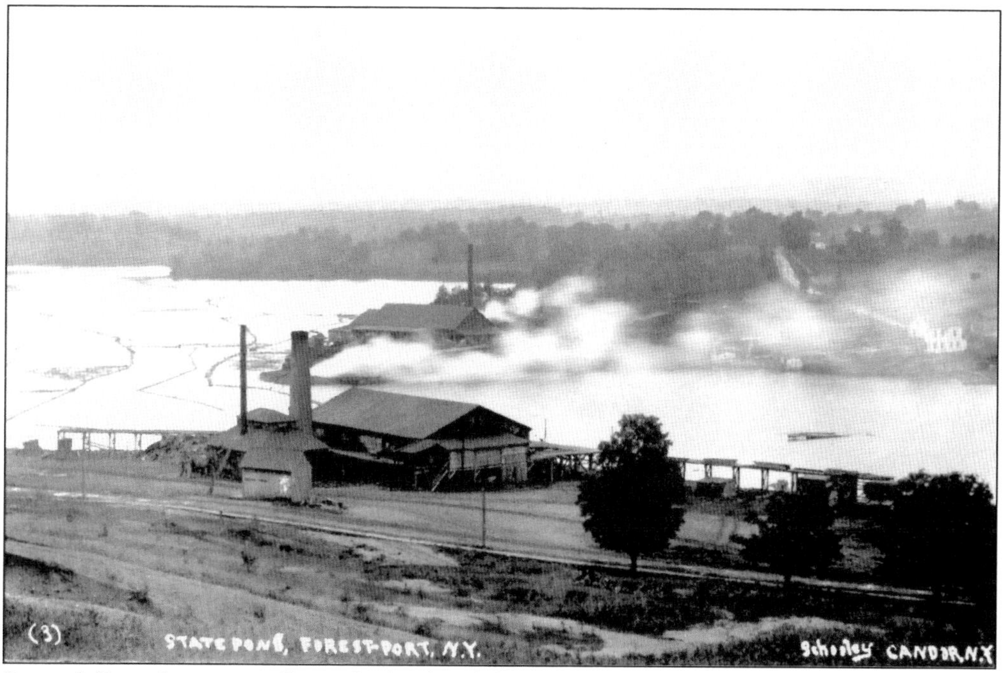


Steam billows from sawmills on the banks of the 150-acre Forestport pond. Many sawmills on Black River converted to steam power around 1860 when the water supply proved unreliable. In the 1870s, the William R. Stamburg mill alone turned out 10 million board feet a year.

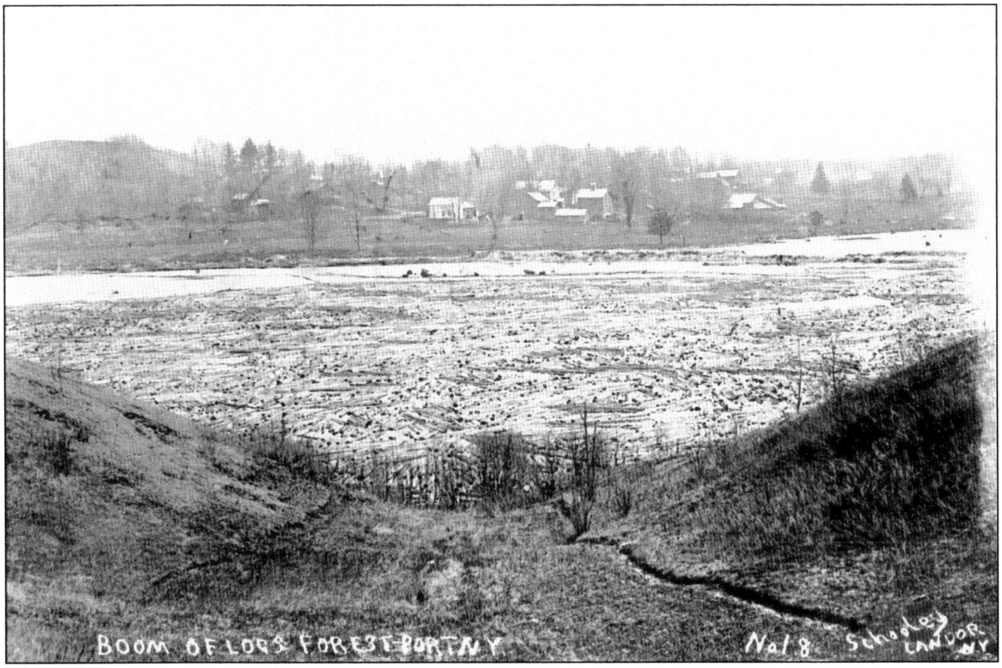


Softwood logs fill Forestport pond. In 1853, Black River was declared a public highway. Log drives did not begin in earnest, however, until 1859. The last drive on Black River was 89 years later, in 1948.

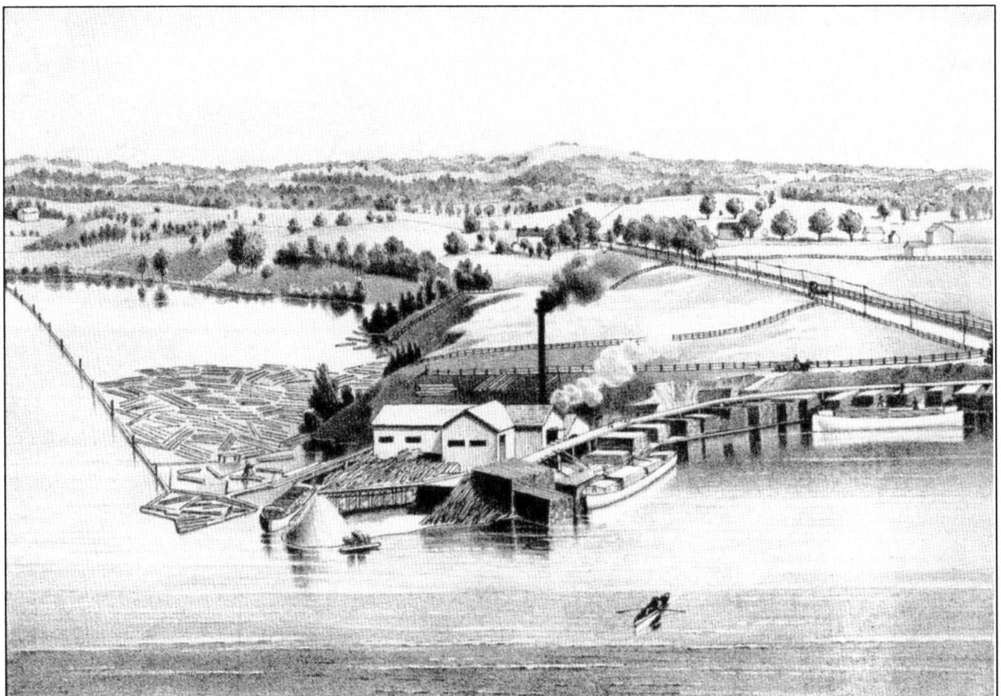

Canal boats take on lumber at a sawmill founded by Civil War veterans Alonzo Denton and Nathaniel Waterbury. In 1868, Forestport had 10 water-powered and 2 steam-powered sawmills.

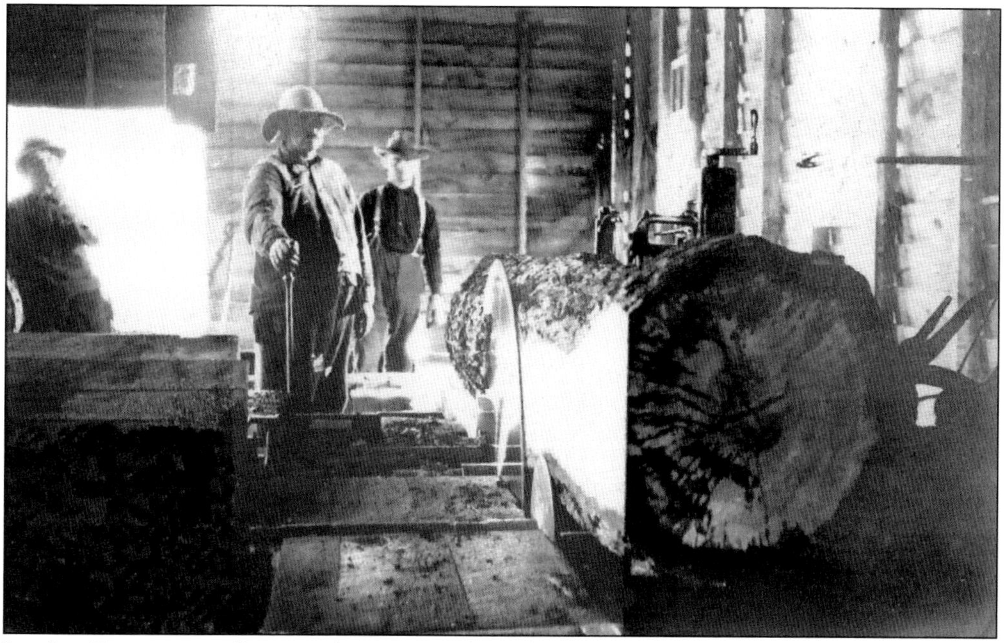

Inside a Forestport mill, a circular saw rips through a massive pine log. Circular saws appeared in the 1850s but were used only in steam-powered mills like this one. (Courtesy of Town of Webb Historical Association.)

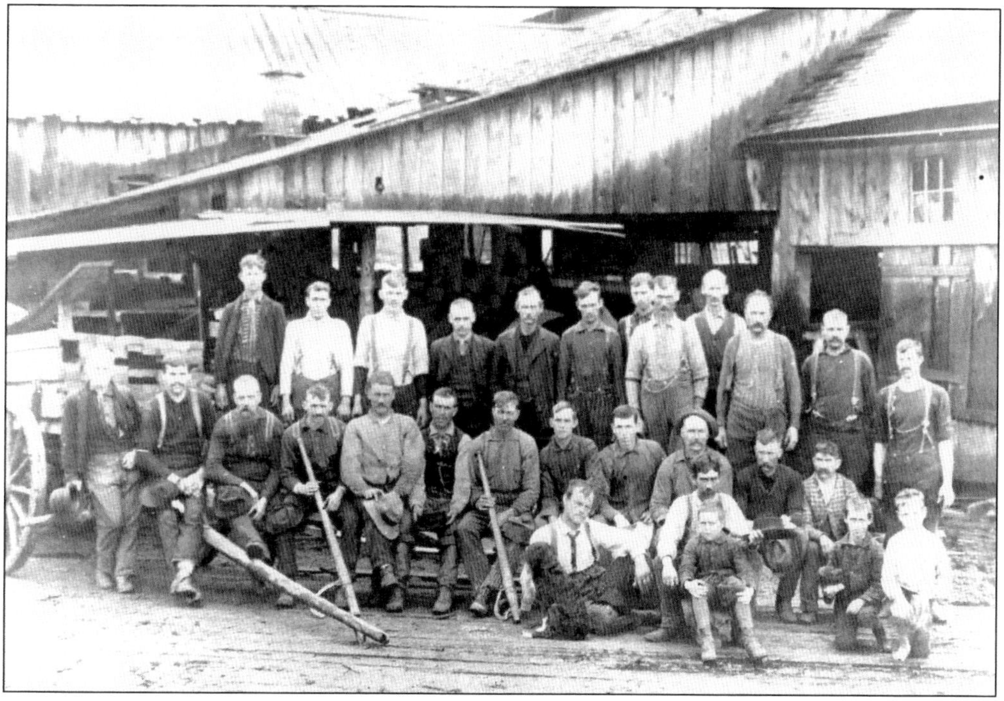

Workers at a Forestport sawmill pose with their peaveys, sons, and mascot. Invented by Maine blacksmith Joseph Peavey in 1858, a peavey is a long wooden staff with a moveable iron hook attached at the bottom. A peavey allows its user to grip logs securely and roll them with leveraged force.

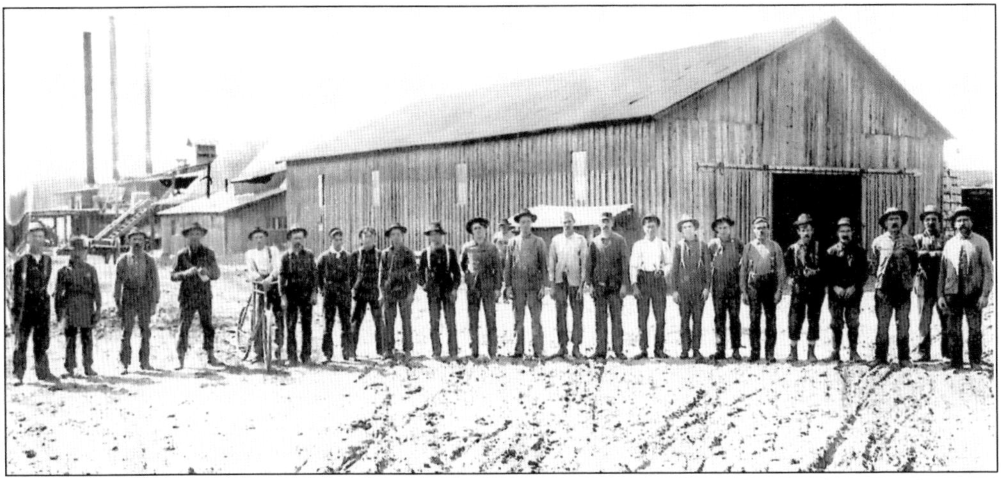

Members of the work force at a Forestport sash and blind factory stand for a photograph.

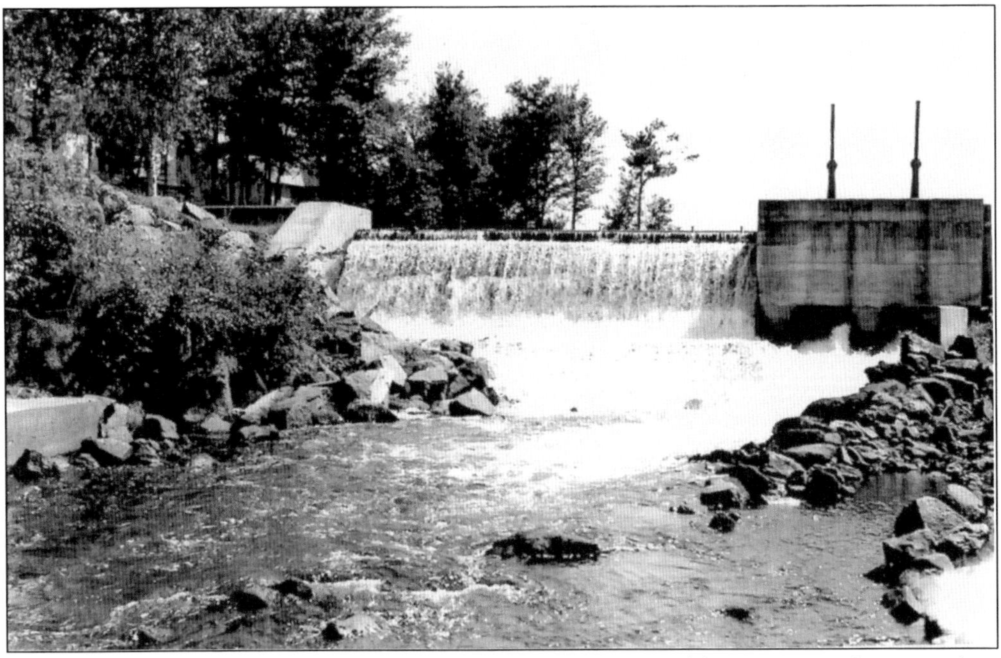

The Old Forge dam was completed in 1881, creating a series of five reservoirs by expanding the existing Fulton Chain of Lakes. Part of the Black River Canal impoundment system, the reservoirs cover approximately 3,500 acres and hold 800 million cubic feet of water. (Courtesy of Town of Webb Historical Association.)

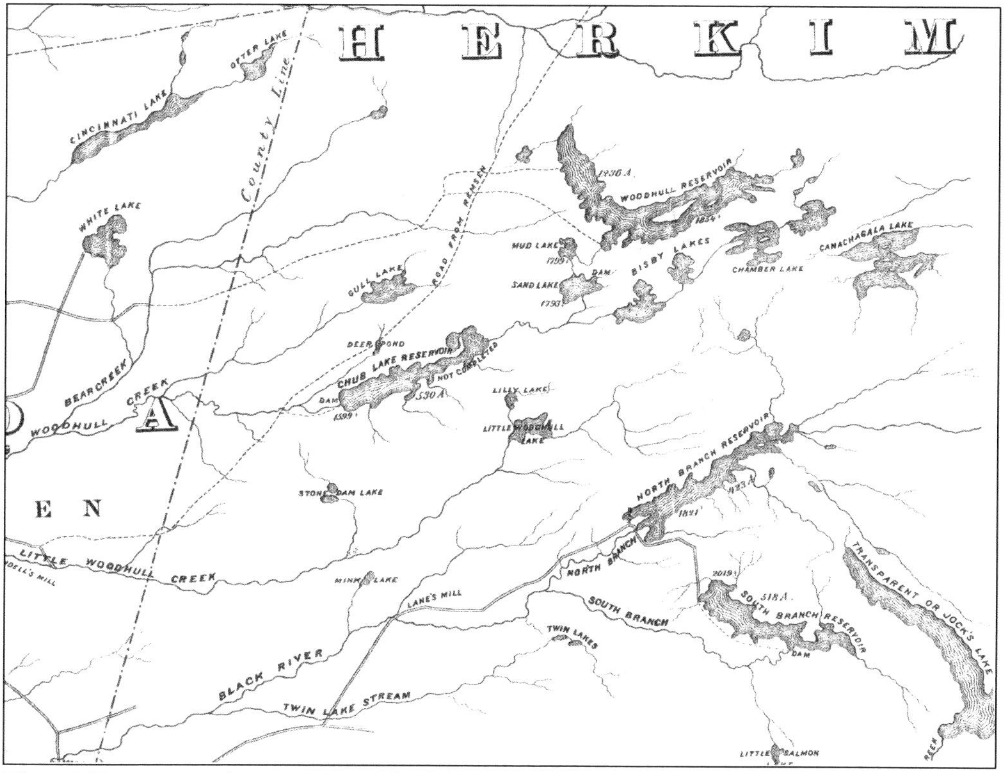


This 1867 state map depicts some of the lakes in Herkimer County, east of Forestport, that were to be included in the Black River Canal impoundment system.

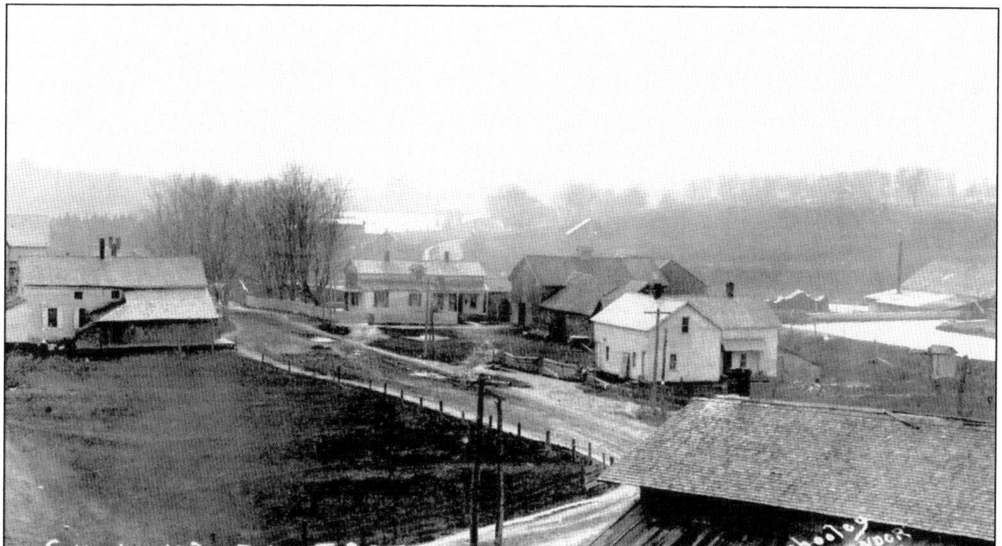


In 1849, a temporary dam was placed on the outlet to Woodhull Lake. By 1885, nearly every river and lake in the region was tapped for water, including North Lake, South Lake, Sand Lake, Twin Lakes, Chub Lake, Canchangala Lake, Bisby Lakes, White Lake, and Moose River. But the reserves were never enough.

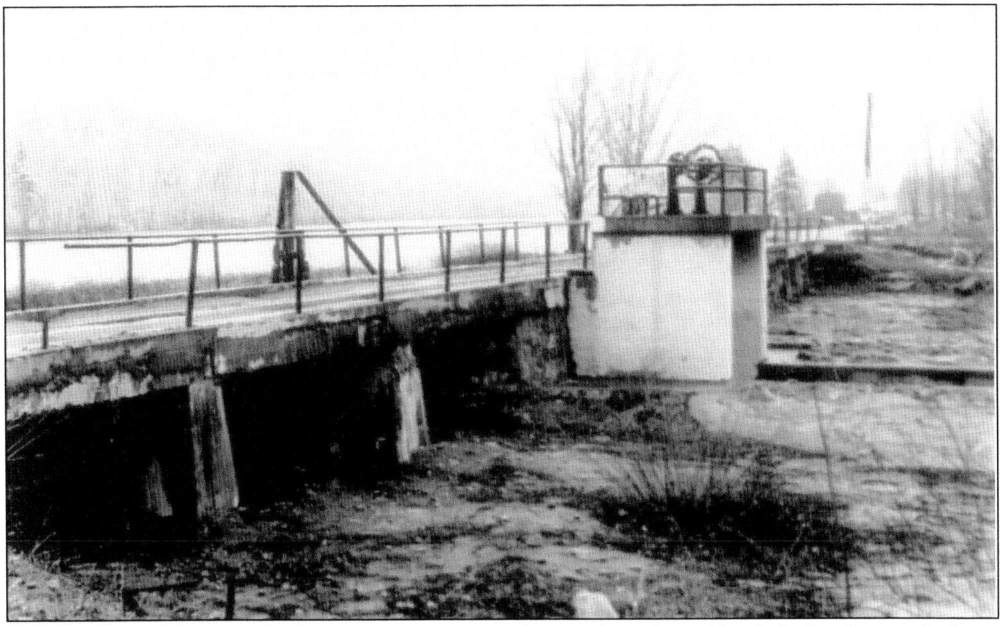

In 1856, the north branch of Black River was dammed to form North Lake. Located about 15 miles from Forestport, North Lake covers about 423 acres and holds 310 million cubic feet of water. In April 1869, an inexperienced gatekeeper at North Lake failed to open the gates following a heavy rain. The dam washed out, causing devastation from Forestport to Watertown. Some 2,400 claims were filed against the state. The photograph dates from the 1960s. (Courtesy of New York State Museum.)

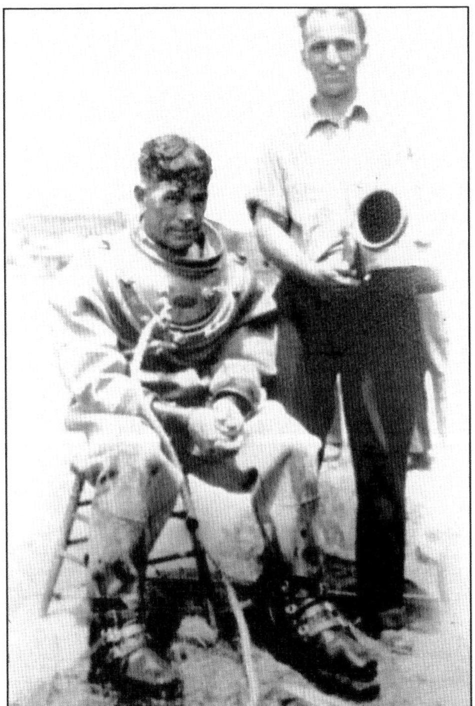

Hard-hat divers were occasionally sent down to clear debris from the gates of North Lake outlet. (Courtesy of New York State Museum.)

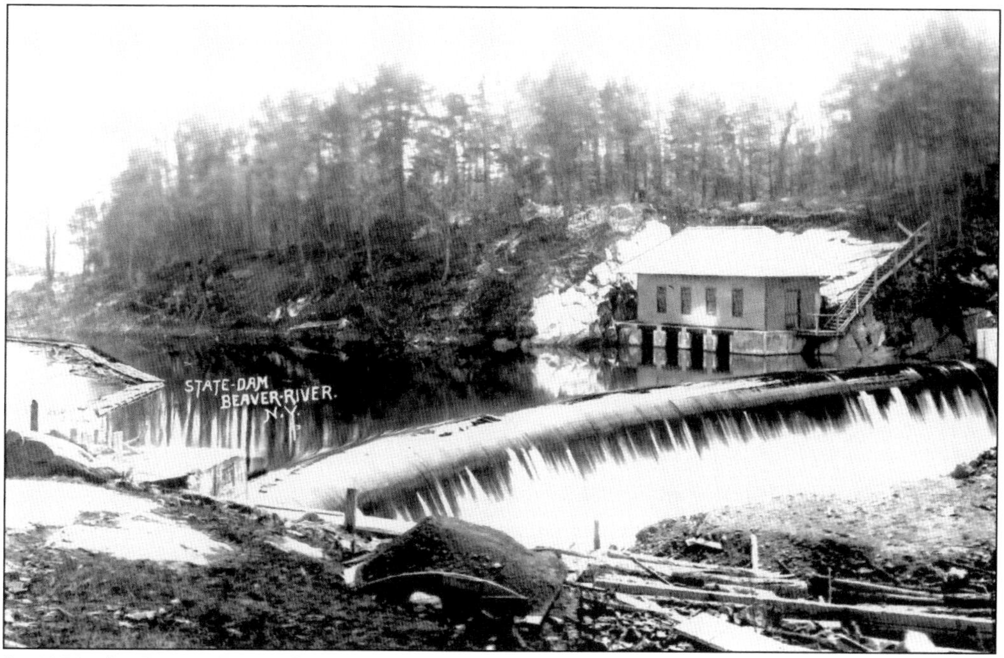


In 1888, the Beaver River dam was completed, creating Stillwater Reservoir in Lewis County, with a capacity of 328 million cubic feet of water. In 1903, a masonry dam was added, and in 1922, the dam was greatly enlarged. (Courtesy of George Davis.)

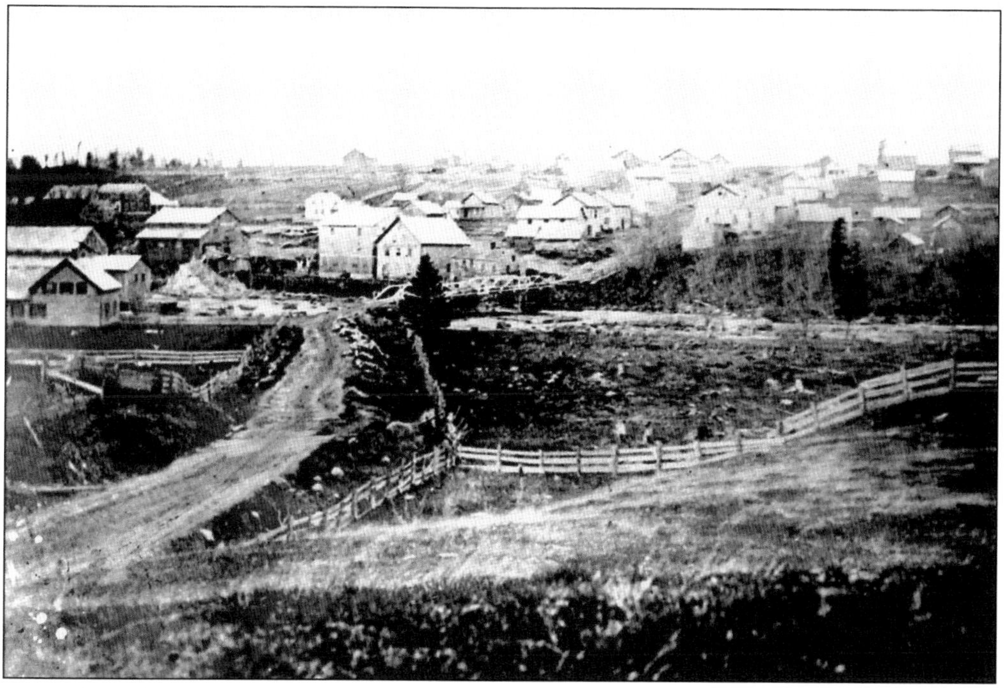

Located on the Erie Canal Feeder two miles east of Boonville, Hawkinsville was a thriving port. William Anderson and Sons' Eureka Tannery shipped hides on the feeder, and William Wenneis's chair factory shipped milled stock.

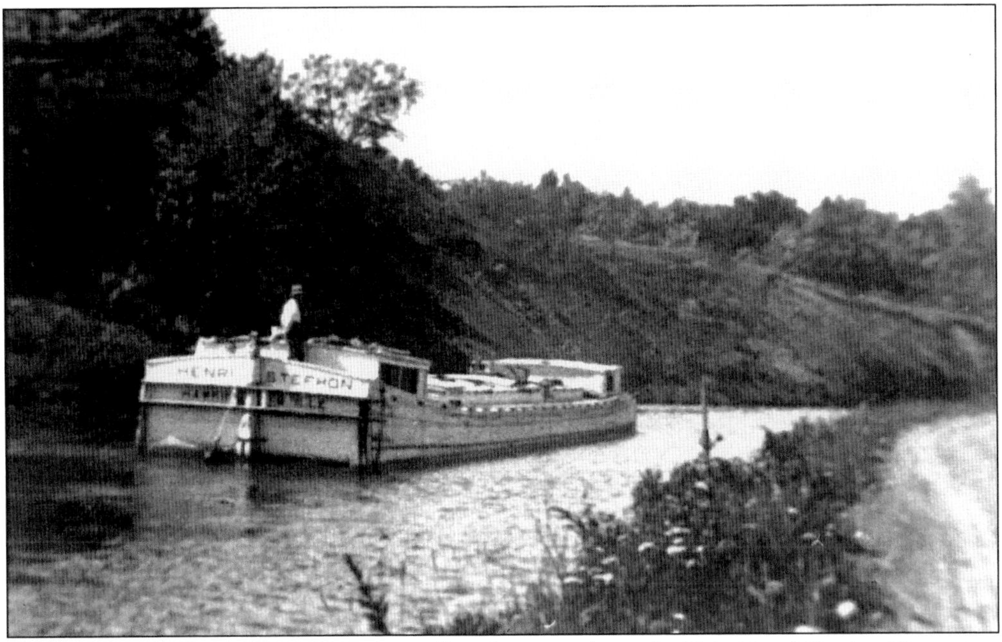


Harry Ford's *Henry Stephon* of Hawkinsville advances along the Erie Canal Feeder. The prism of the feeder was somewhat larger than that of the Black River Canal, measuring 46 feet at the surface and 30 feet at the bottom. The bottom of the feeder, like that of the canal, was puddled, covered with a clay slip to make it less permeable. (Courtesy of New York State Museum.)

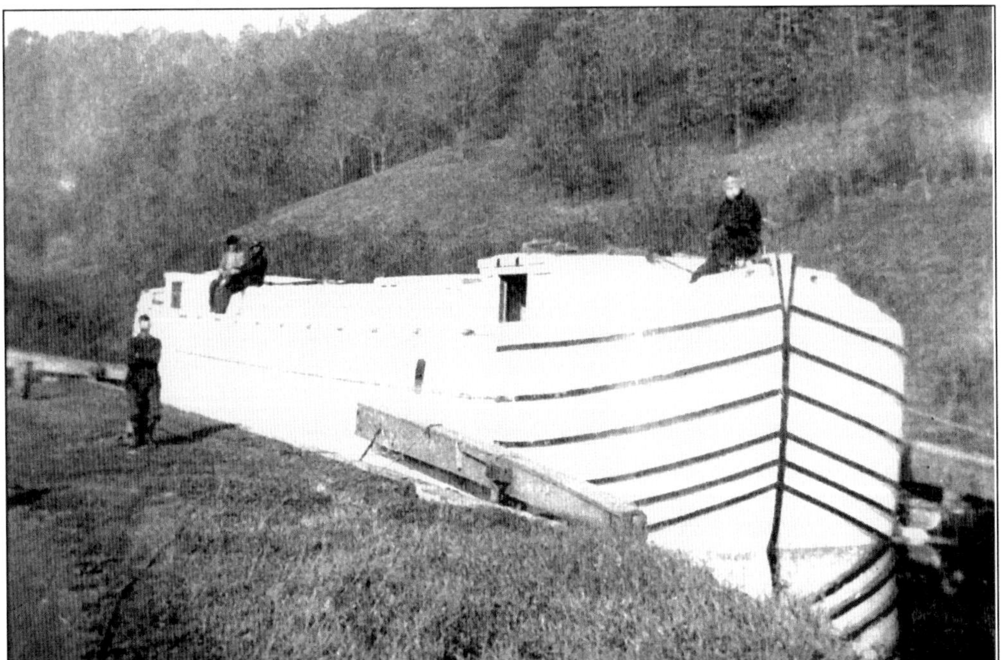

Owned by Tom Shanks and named for his two sons, the *Harrison and Kenneth*, a Durhamville laker, heads north through the Lansing Kill Gorge. Tom Shanks shipped potatoes purchased from Sam Dorn, the "Potato King" of Jackson Hill. (Courtesy of New York State Museum.)

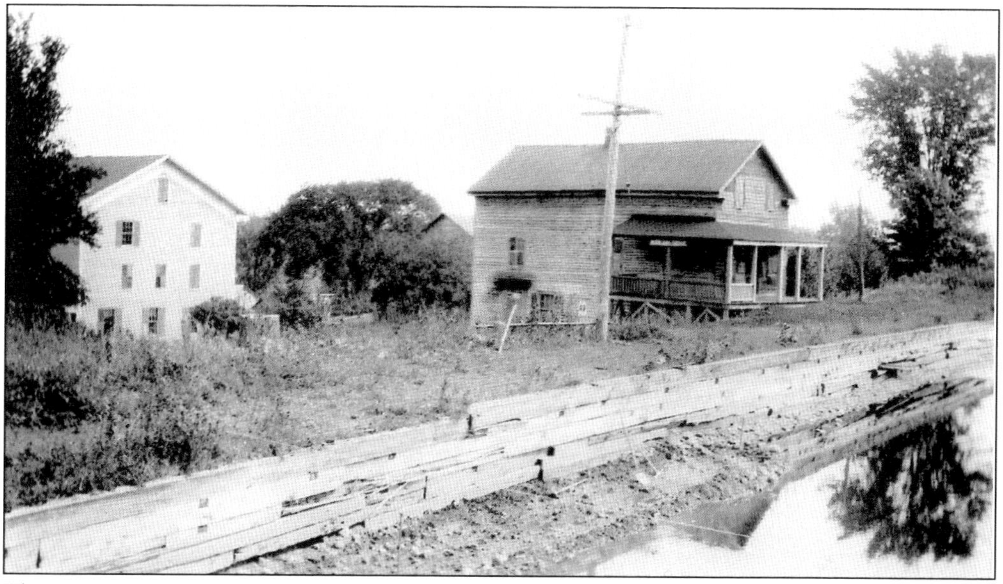

The Erie Canal Feeder runs past the Yager family farm, located between Forestport and Hawkinsville. Timbers on the berm side form a retaining wall to halt erosion of the sandy soil. The photograph was taken in the 1960s.

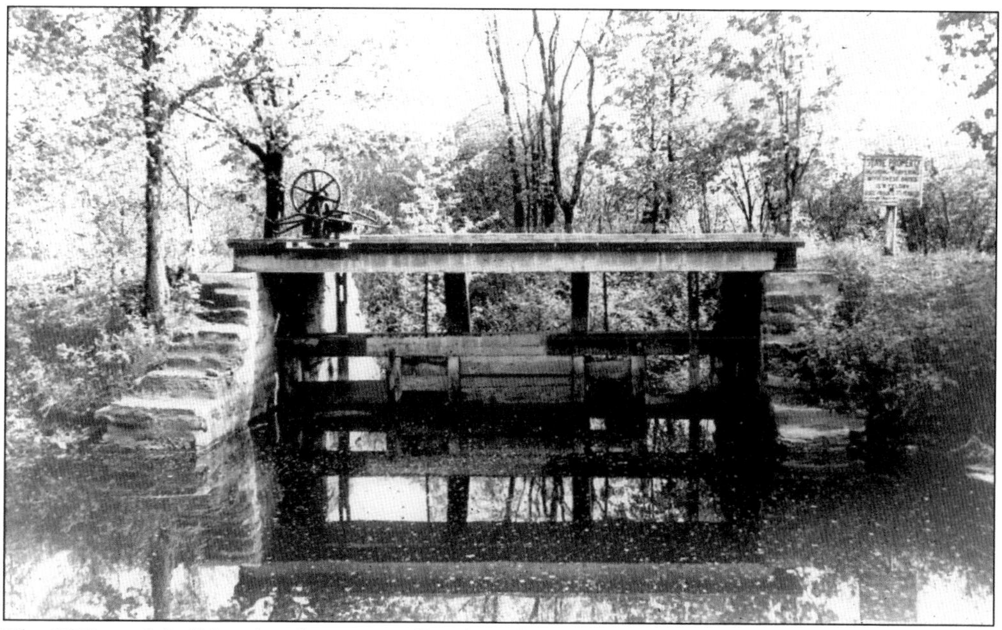

The Williams's bulkhead and gate near Dutch Hill Road in Forestport allows excess water to be released from the Erie Canal Feeder. The photograph dates from the 1960s. (Courtesy of New York State Museum.)

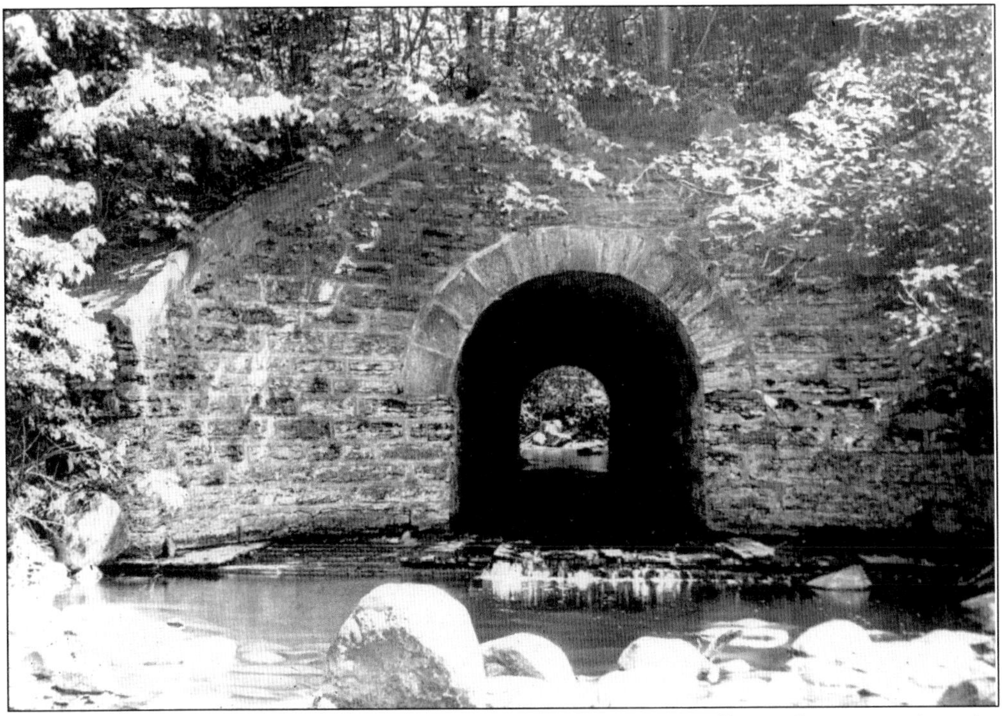

One of three on the Erie Canal Feeder, this culvert serves as a small aqueduct, extending the canal across a stream. (Courtesy of New York State Museum.)

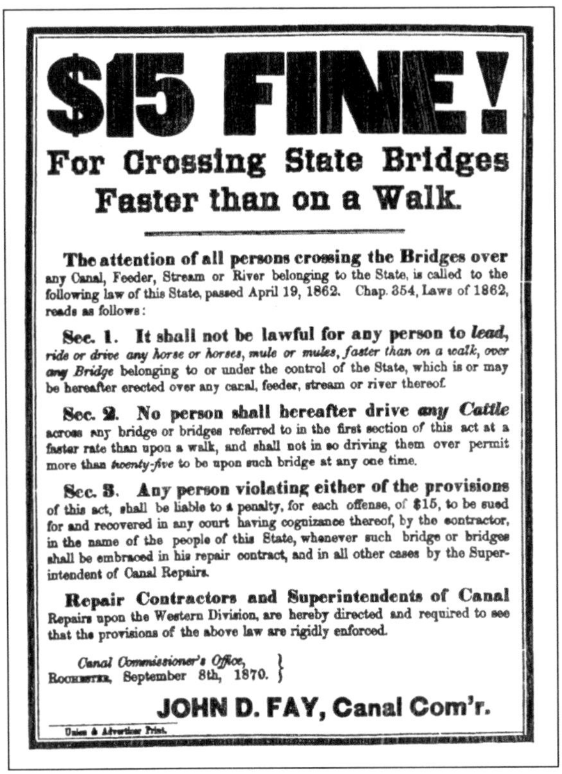

# $15 FINE!

## For Crossing State Bridges Faster than on a Walk.

**The attention of all persons crossing the Bridges over** any Canal, Feeder, Stream or River belonging to the State, is called to the following law of this State, passed April 19, 1862. Chap. 354, Laws of 1862, reads as follows:

**Sec. 1. It shall not be lawful for any person to** ***lead,*** *ride or drive any horse or horses, mule or mules, faster than on a walk, over any Bridge* belonging to or under the control of the State, which is or may be hereafter erected over any canal, feeder, stream or river thereof.

**Sec. 2. No person shall hereafter drive** ***any Cattle*** across any bridge or bridges referred to in the first section of this act at a faster rate than upon a walk, and shall not in so driving them over permit more than *twenty-five* to be upon such bridge at any one time.

**Sec. 3. Any person violating either of the provisions** of this act, shall be liable to a penalty, for each offense, of $15, to be sued for and recovered in any court having cognizance thereof, by the contractor, in the name of the people of this State, whenever such bridge or bridges shall be embraced in his repair contract, and in all other cases by the Superintendent of Canal Repairs.

**Repair Contractors and Superintendents of Canal** Repairs upon the Western Division, are hereby directed and required to see that the provisions of the above law are rigidly enforced.

*Canal Commissioner's Office,*
Rochester, September 8th, 1870.

**JOHN D. FAY, Canal Com'r.**

Union & Advertiser Print.

Citing an 1862 law, this broadside warned citizens that they risked a $15 fine if they rode or drove animals over a canal bridge any faster than a walking pace. A total of 14 farm and road bridges spanned the Erie Canal Feeder.

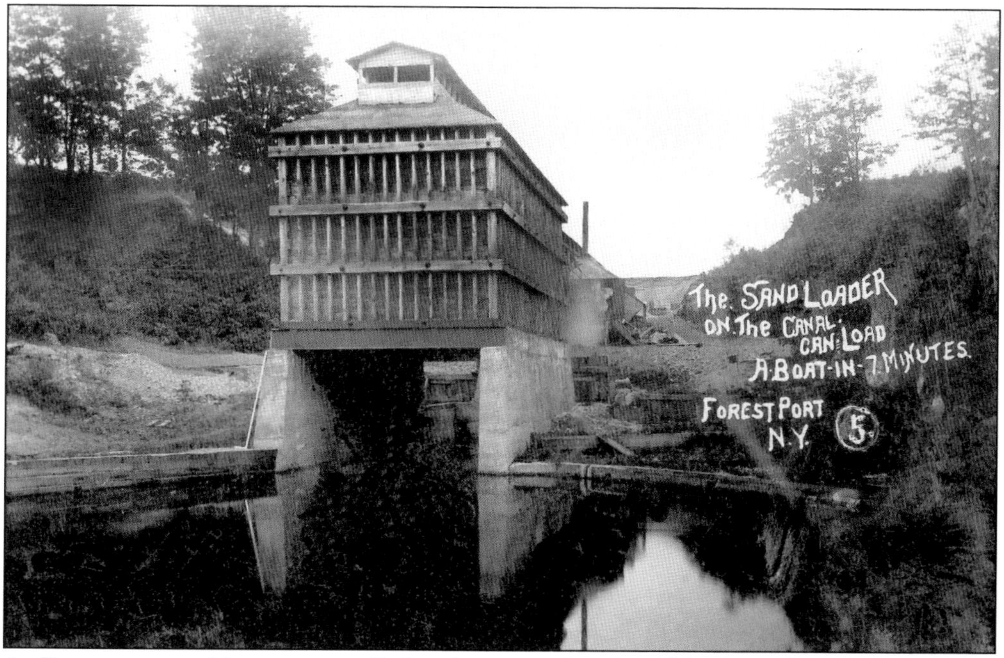


At Forestport, canal boats were pulled under an overhead sand loader and were quickly filled with the alluvial sand that abounds in the area. (Todd Weseloh Collection.)

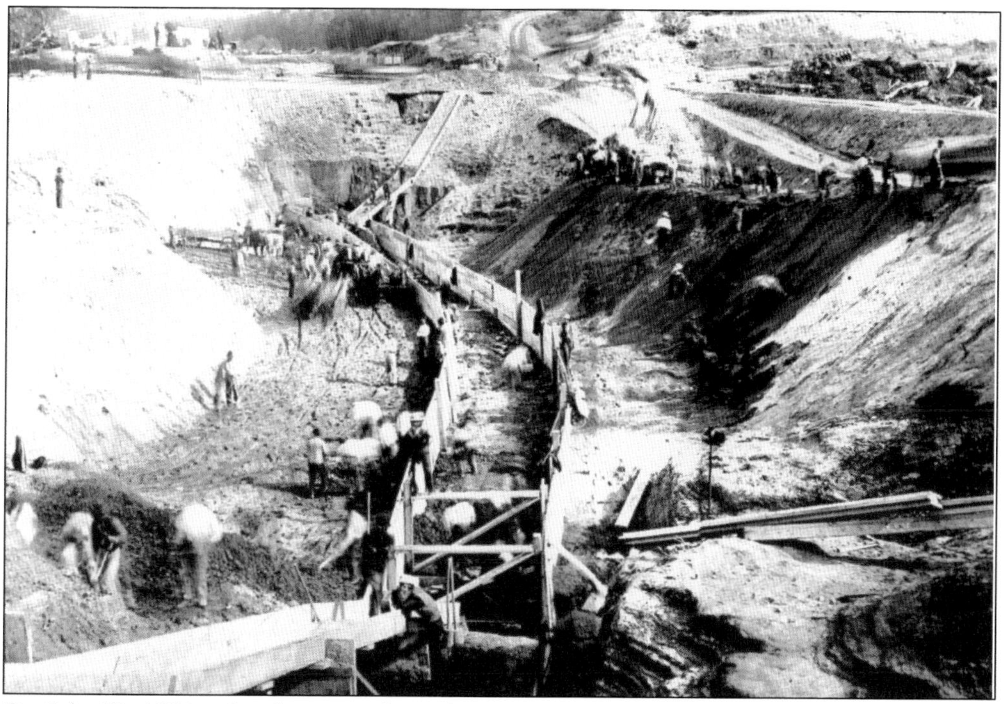

On July 17, 1897, a break occurred on the Erie Canal Feeder near Forestport, washing out 400 feet of the towpath bank and 50 feet of the feeder bottom. Repair crews built two rows of sheet piling, each 53 feet high. The core was then filled with puddled sand and gravel.

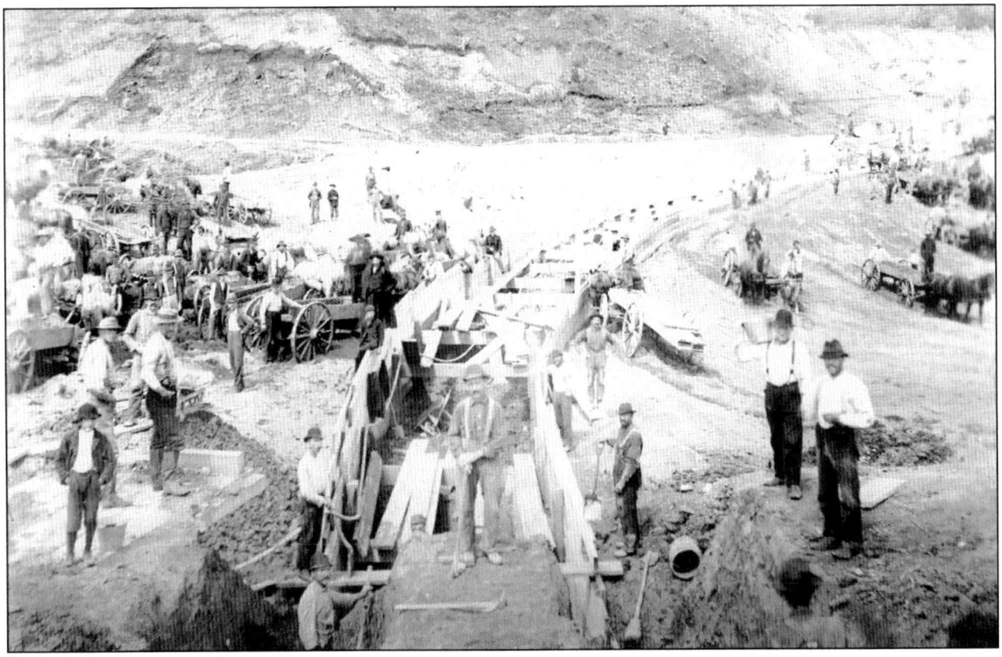

The state hired 1,700 men to repair the 1897 break on the Erie Canal Feeder, with each man earning $1.65 a day. In addition, 250 teams of horses were recruited, with each driver earning 35¢ an hour. The project took a month and cost $63,000.

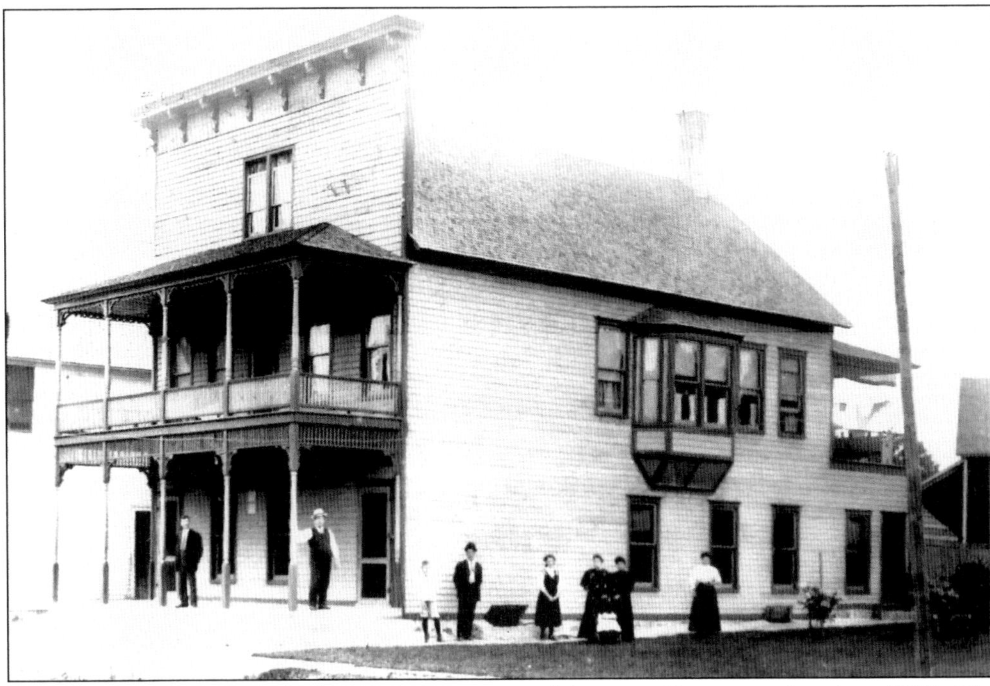

Rebuilding after a fire, Charley O'Connor opened his new Forestport hotel in 1899. The hotel had 12 rooms, a ground-floor bar, and a dining room. Guests paid $1.50 per night for room and board.

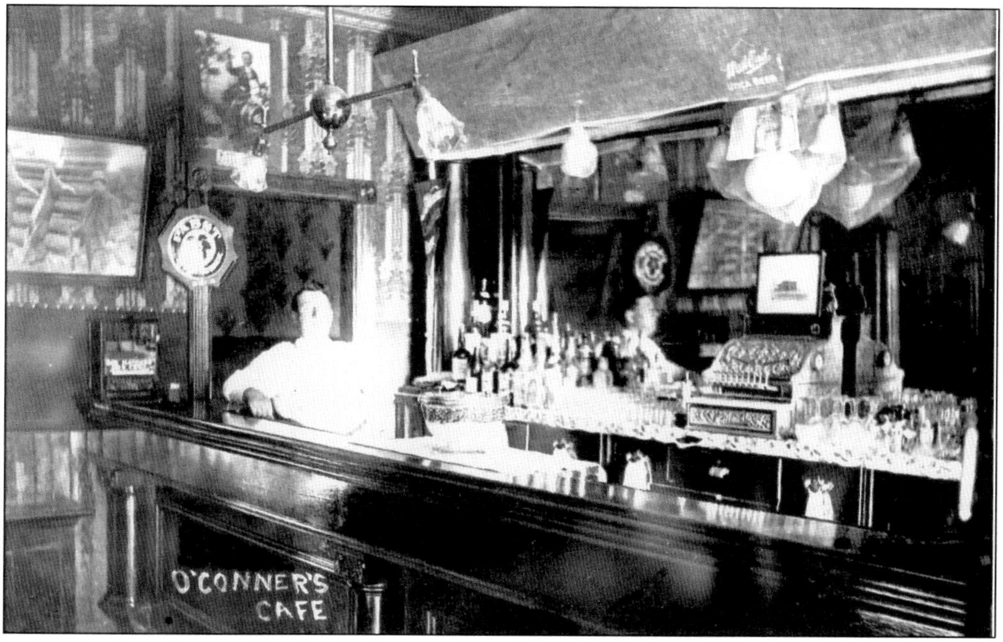


In 1898, saloon keeper Charley O'Connor allegedly offered money to anyone who would cause a break in the Erie Canal Feeder.

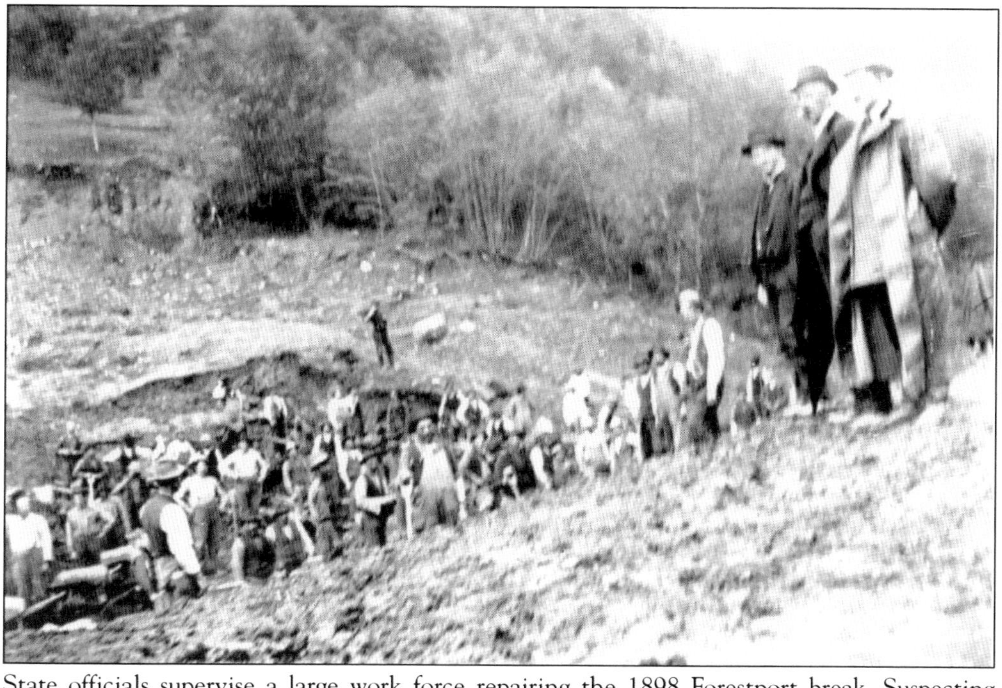

State officials supervise a large work force repairing the 1898 Forestport break. Suspecting sabotage, state superintendent John Partridge hired the Pinkerton Detective Agency to investigate the washout. Agents discovered that saloon keepers had promised to pay men to undermine the towpath wall, hoping that an influx of workers to the area would boost business at their establishments. In the end, eight men, including Charley O'Connor, were punished. A third break occurred in 1899.

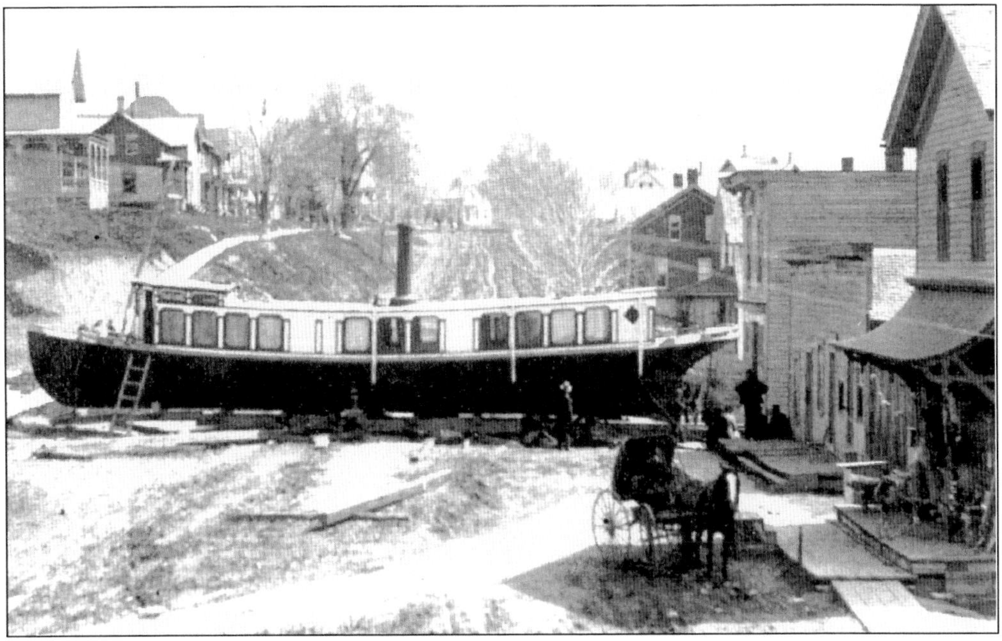

A packet steamer has been hauled ashore for repairs in Forestport. In the 1890s, Ike Scouten's steamer, the *Ollie*, provided daily passenger service between Forestport and Boonville, departing Forestport at 9:00 a.m., arriving in Boonville at noon, and leaving at 5:00 p.m. for the return trip to Hawkinsville. (Courtesy of Erie Canal Museum.)

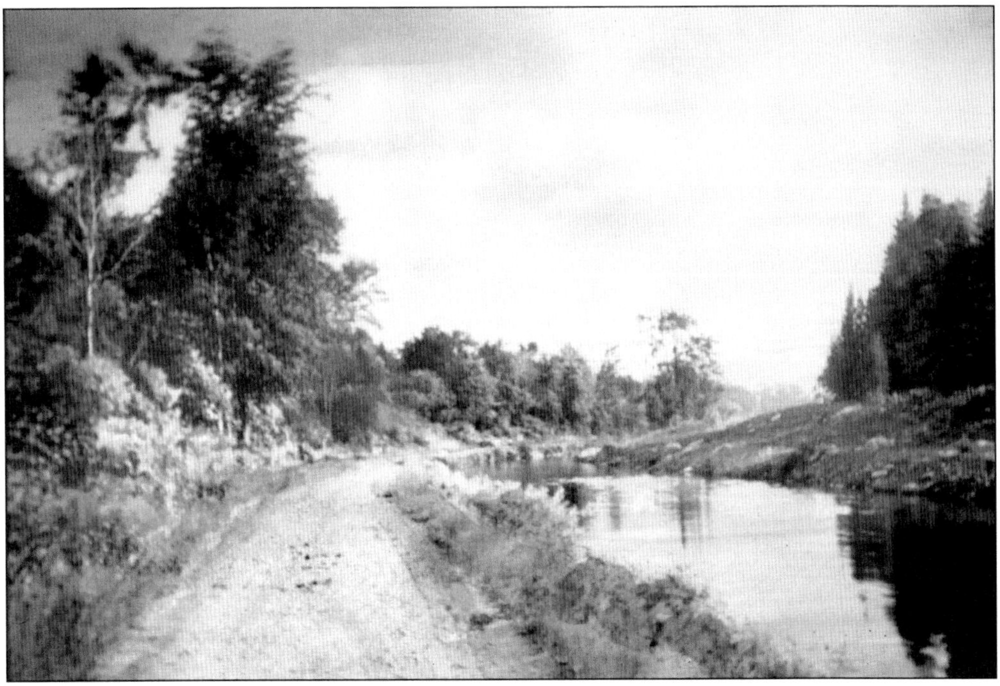

The feeder canal looks much the same today as it did in the 1920s when Federick B. Hough took this photograph.

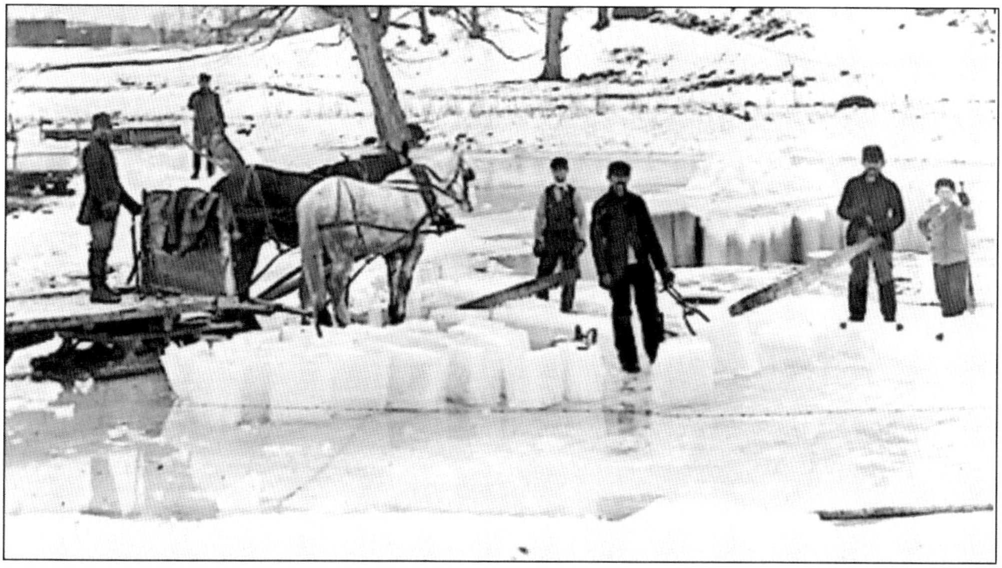

At Forestport, Black River ice was cut in winter and stored until spring, when it was transported by canal to the Knickerbocker Ice Company in New York City.

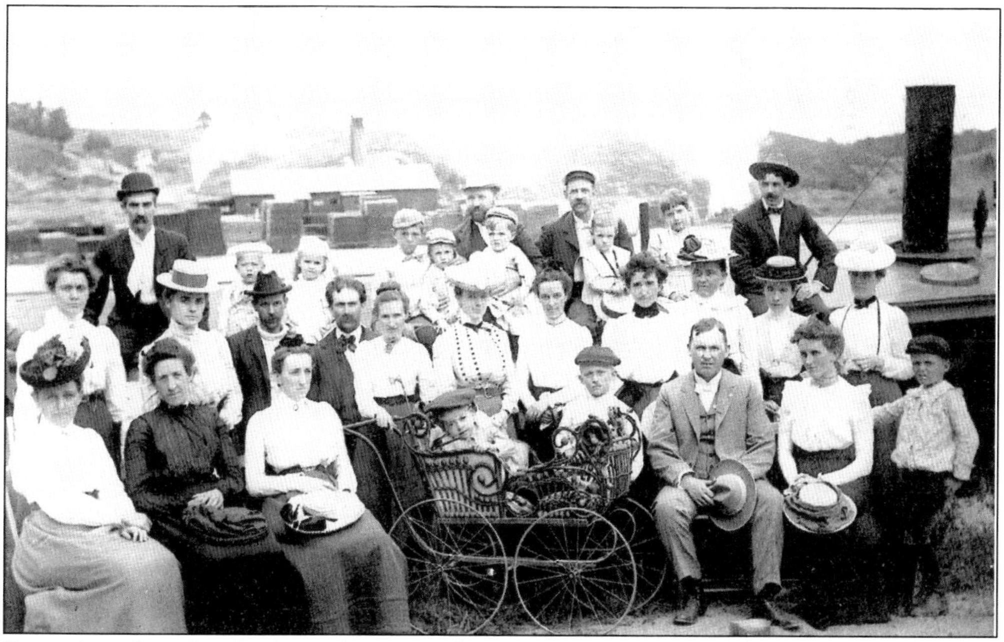

Dressed in their Sunday best, this group from Boonville is on an outing to Forestport. The small steamer on the right may be the *Ollie*. In the background is Forestport pond, with its busy sawmills.

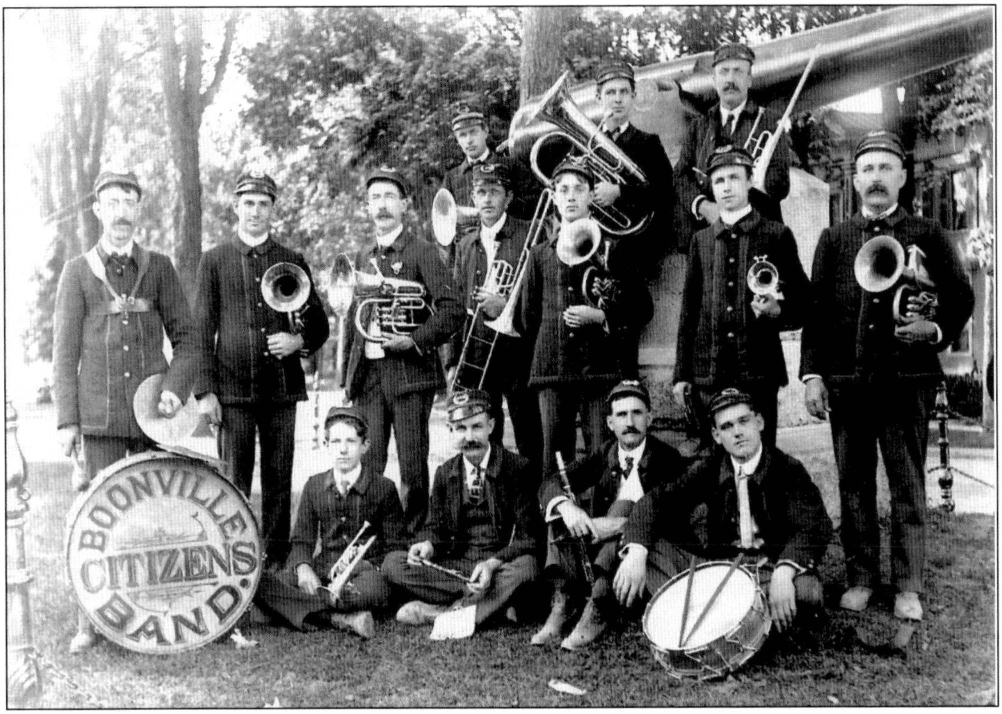


The Boonville Citizens Band was one of several bands hired to play at social gatherings and political rallies held along the Erie Canal Feeder. The scene here is the village park in Boonville.

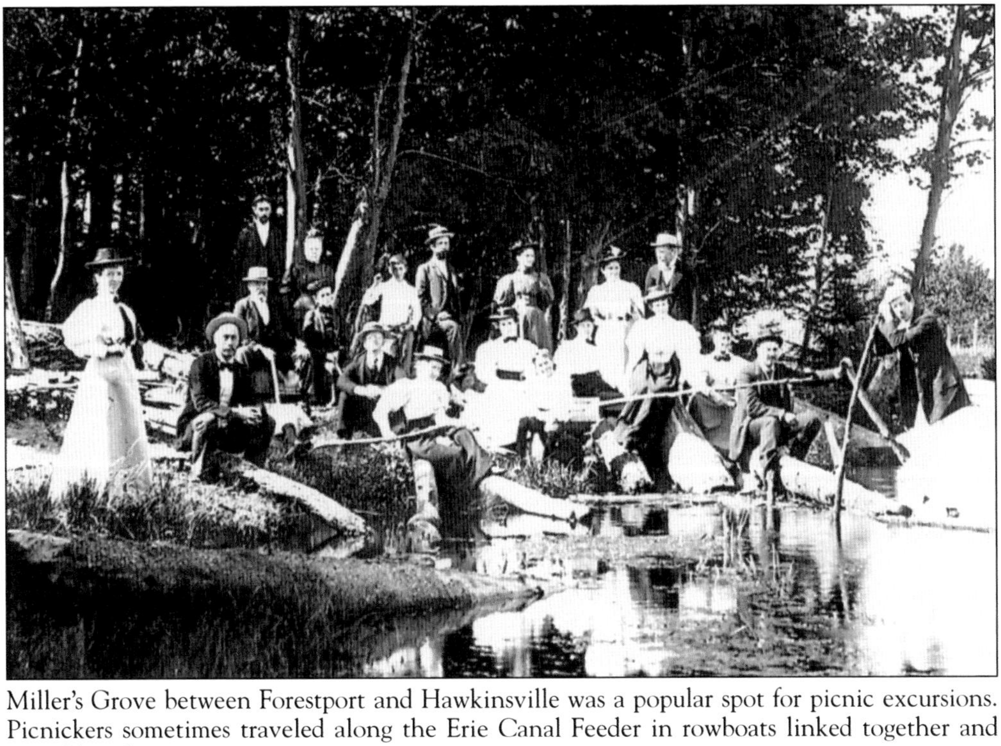

Miller's Grove between Forestport and Hawkinsville was a popular spot for picnic excursions. Picnickers sometimes traveled along the Erie Canal Feeder in rowboats linked together and towed by a horse.

# Three

# THE GREAT HARVEST

The canal's arrival made it economically feasible to harvest softwood in the western Adirondacks. Lumber camps and sawmills sprang up along the Black River and its tributaries, and logging quickly became the region's dominant industry. Timber, sawn lumber, and other wood products (pulpwood, firewood, shingles, staves, spruce gum, and pitch) were by far the most important commodities shipped on the canal. Virgin spruce from the area was instrumental in building the port of New York City. Giant spars, often more than 70 feet long, were towed through the Black River Canal on multiple cribs. These water-resistant spars became the masts of sailing ships and the pilings of New York City's wharves and docks. One important by-product of the lumber industry was hemlock bark. Much of the bark was used by local tanneries, but some was shipped to tanneries in New England, where hemlock forests had largely disappeared. Since only softwood could be floated downstream to mills, Adirondack hardwoods were left largely untouched.

Between 1855 and 1890, forest products accounted for well over 80 percent of canal tonnage. In 1866, some 29 million board feet of lumber, 135,000 cubic feet of timber, and 10,000 cords of wood were shipped on the canal, destined primarily for Albany's lumber district. In the following year, the Utica and Black River Railroad was extended north from Boonville to Lyons Falls, and the railroad began to compete with the canal in hauling forest products. The canal had an important advantage, however. Since the Adirondack forests were on the eastern side of the canalized Black River, boats could be loaded directly from the riverbank. Moreover, canal boats were able to pick up loads at the great mills on Moose River and Beaver River. The canal retained its dominance in hauling freight until regional softwoods were no longer available.

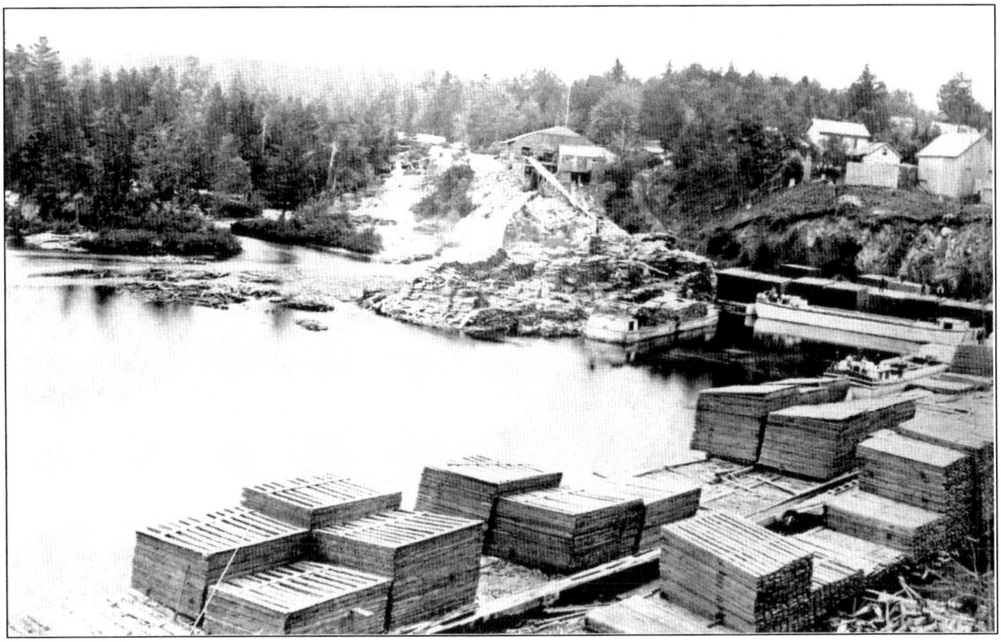

The Black River Canal opened up an estimated 900,000 acres of Adirondack forest to logging. Here, lumber is being loaded on boats at Gouldtown, about one mile upstream from the mouth of the Moose River. The year is 1896.

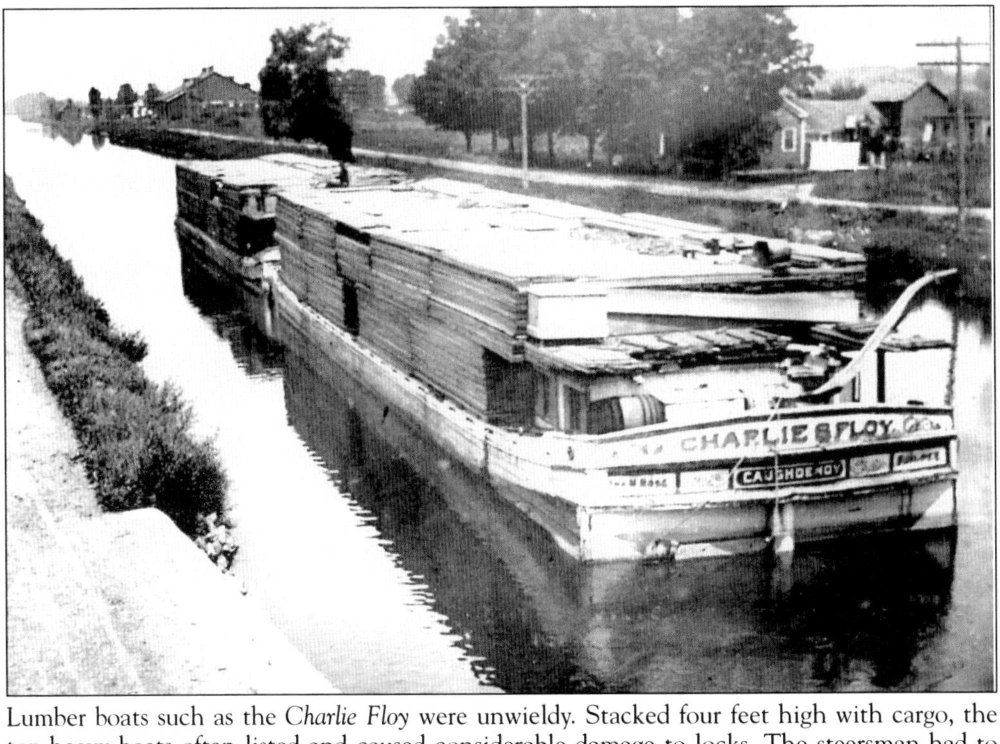


Lumber boats such as the *Charlie Floy* were unwieldy. Stacked four feet high with cargo, the top-heavy boats often listed and caused considerable damage to locks. The steersman had to stand on an elevated platform to see across the cargo, and he needed a long-handled tiller to steer the boat. The scene is near Durhamville on the Erie Canal.

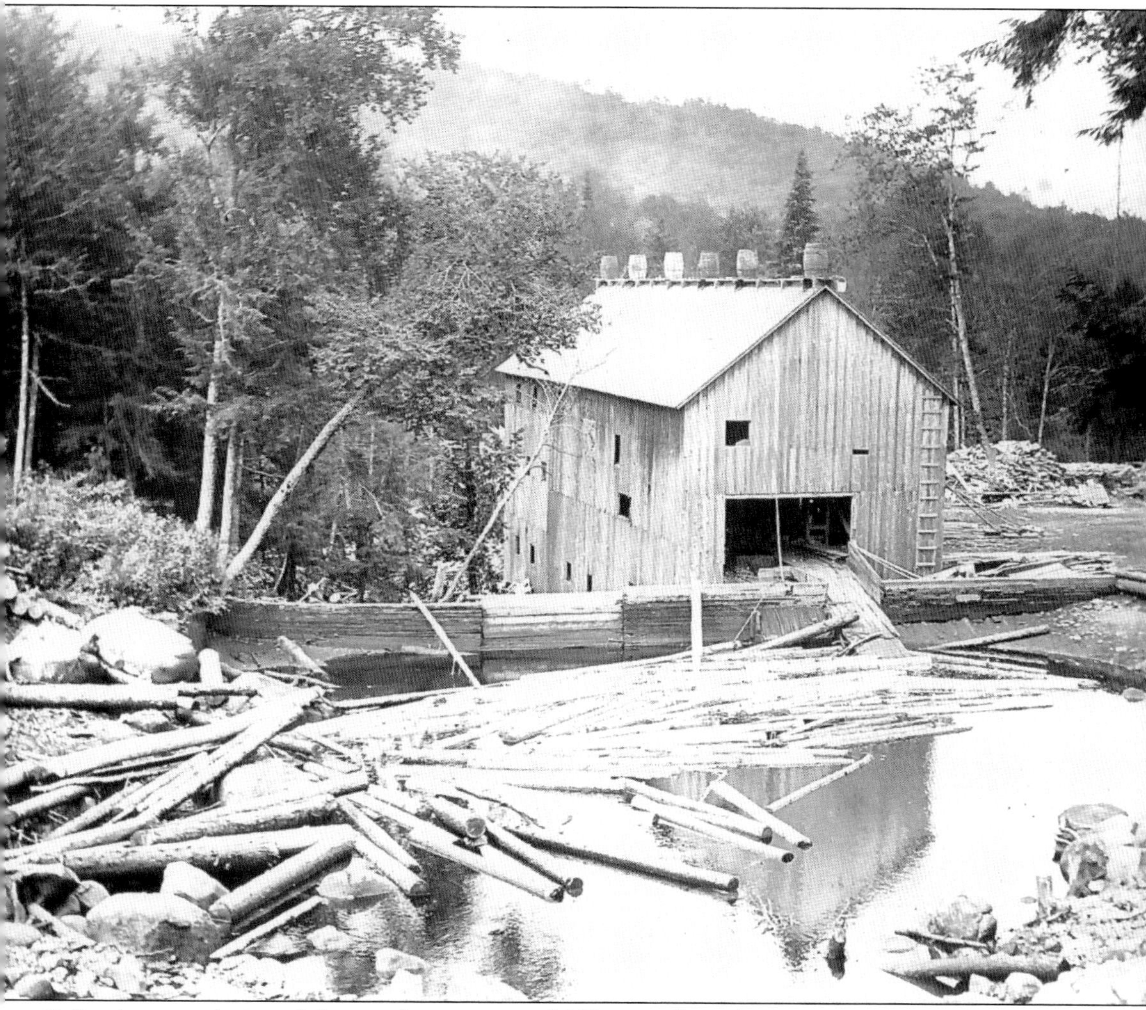

Following completion of the canal, an estimated 200 sawmills and 40 tanneries materialized in the Black River watershed. The mill pictured here is at Fifth Lake. (Courtesy of Town of Webb Historical Association.)

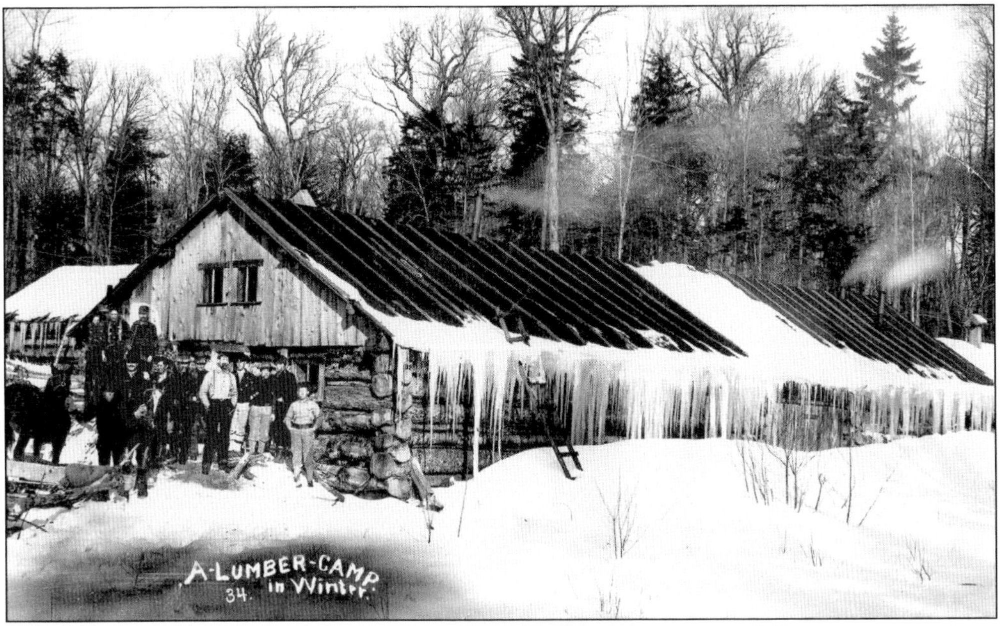


Logging camps were originally built within a mile of a major stream. When trees near these streams disappeared in the late 1860s, camps began to move deeper into the woods. Well-placed camps were within a mile's walk of all cutting and were used for three or more years.

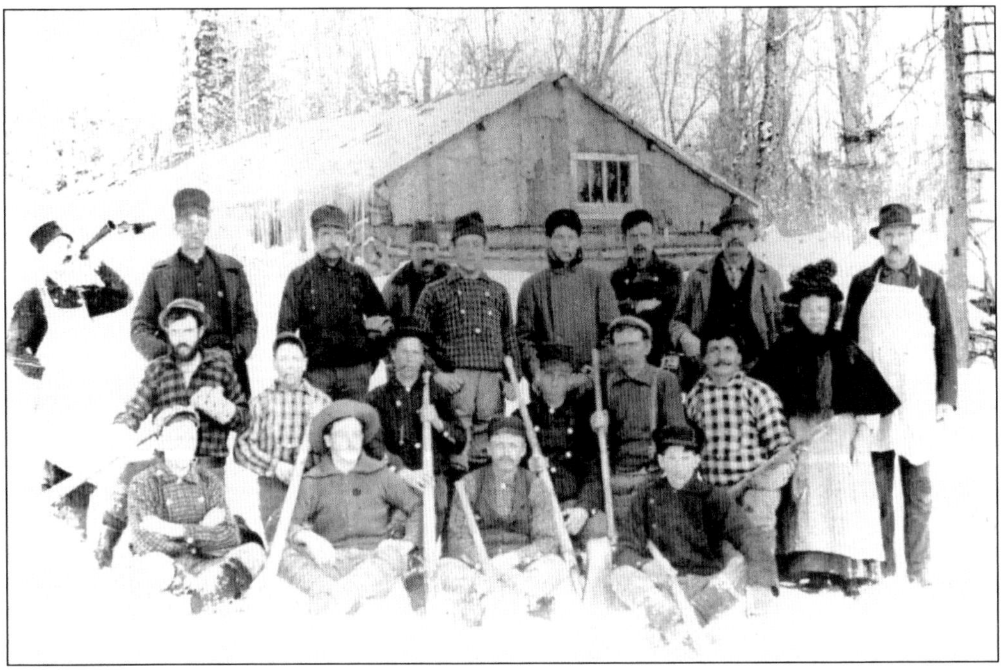

A logging camp typically housed 15 to 45 men. In autumn, small crews would build a camp that included a bunkhouse, cookhouse with dining area, blacksmith shop, small office, and horse barns. In the mid-1870s, loggers earned about $25 to $30 a month.

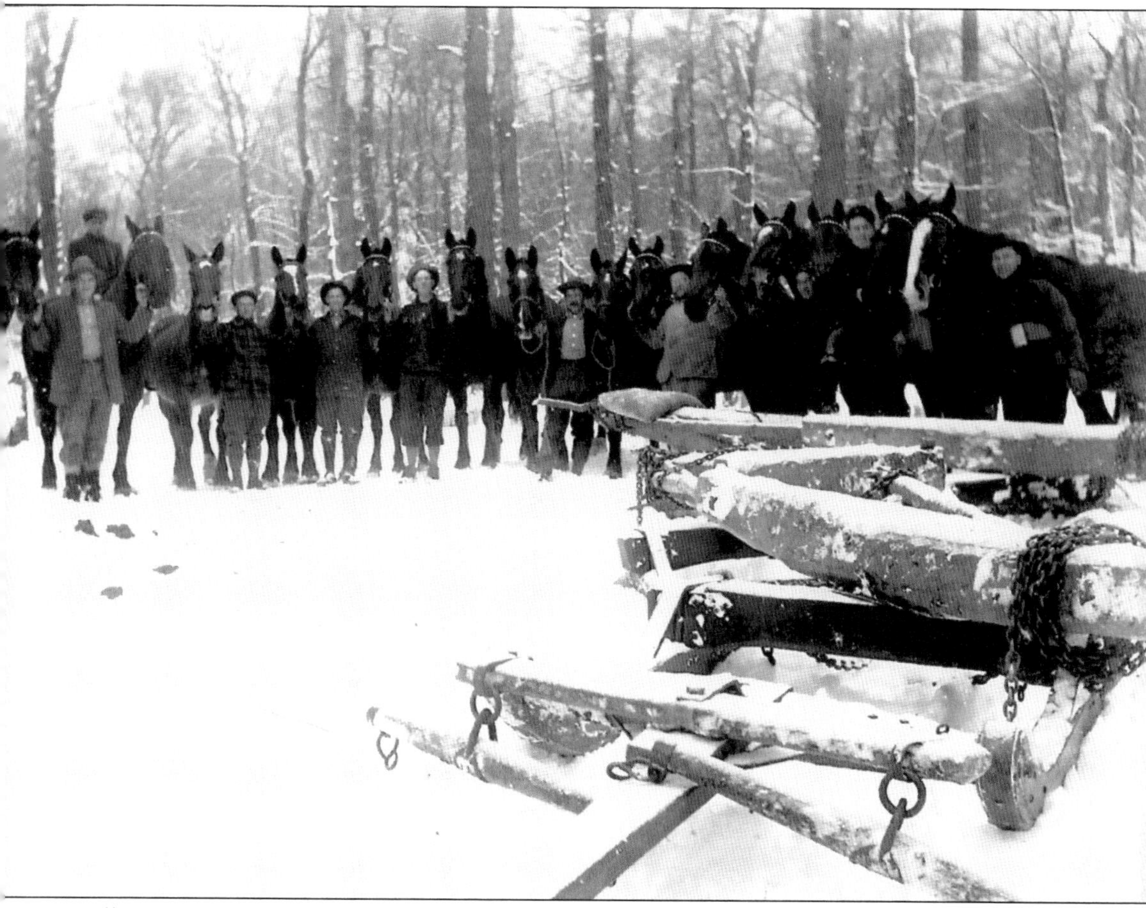

Canallers often hired out their horses for logging in the off-season. Logs were typically bobbed over iced roads on sleds equipped with 8-foot runners and 10-foot bunks, or platforms. On relatively level terrain, two or more sleds might be linked together.

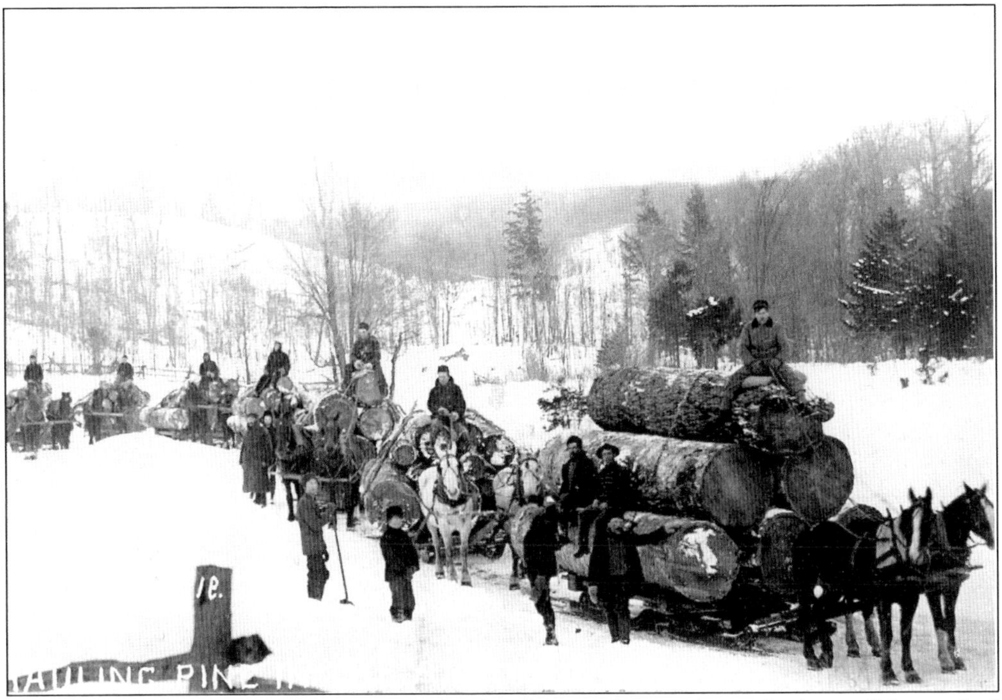

Horses are hauling virgin pine logs to the banks of Moose River. There, the logs will be stacked at landings and floated downstream with the spring thaw. White pine was highly desirable, but it constituted only about three percent of the original forest and was the first species to vanish in large quantities

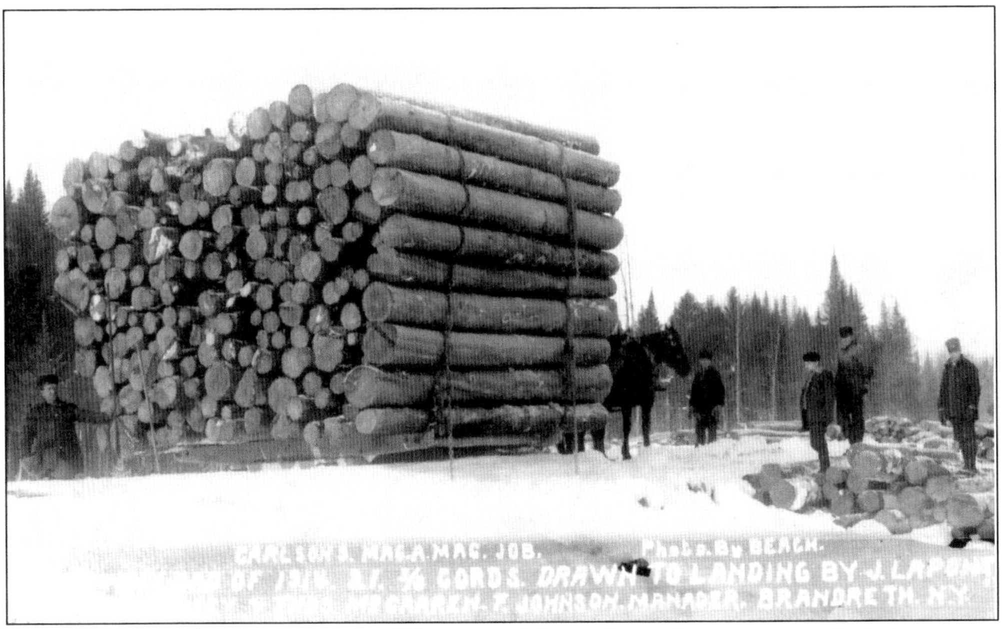


Spruce constituted only 15 percent of the original forest, but it became increasingly important as white pine disappeared. Initially, only 12-inch-diameter logs were harvested, but in later harvests, even 5-inch logs were cut for the paper and pulp industry.

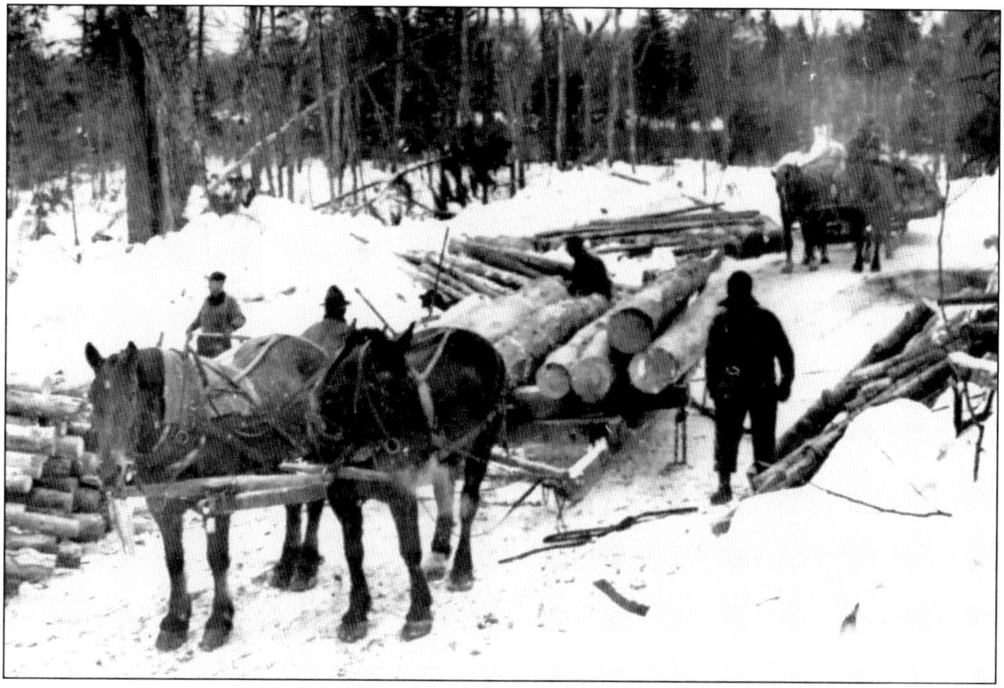

Spruce was cut into standard 13-foot 4-inch logs called markets. In 1901, the average spruce was 70 feet high with a clear trunk of 46 feet and yielded about 3.5 market logs. Spruce bark was often peeled off logs to make them more buoyant. This was done in late spring or early summer when the trees were growing.

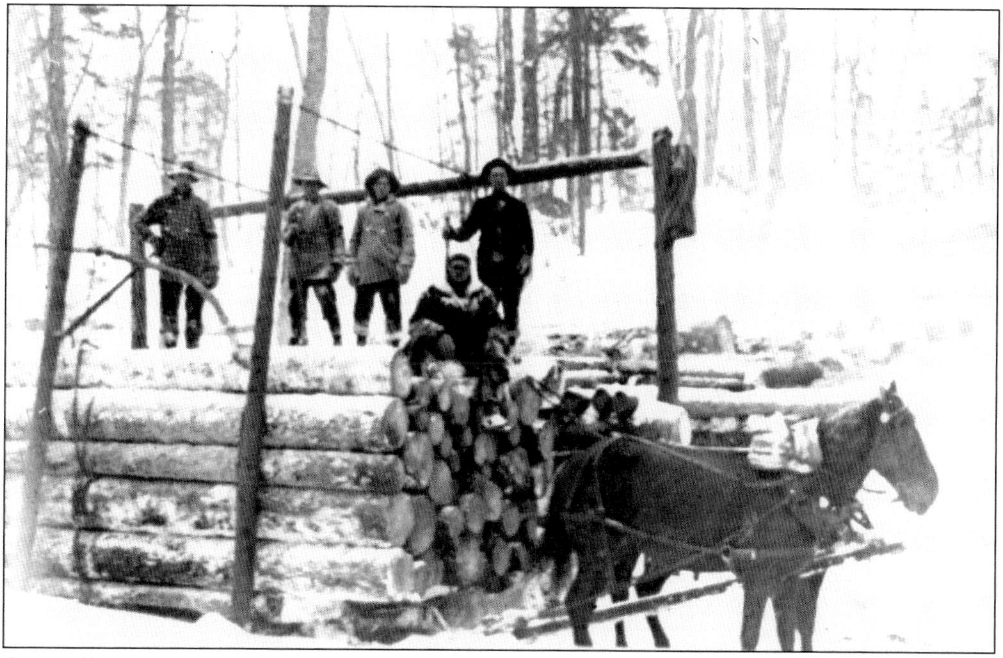

Once all merchantable spruce had been harvested, lumber companies frequently abandoned their land rather than pay property taxes. Through tax sales, the state acquired large tracts that formed the core of what would, in 1885, become the Adirondack Forest Preserve.

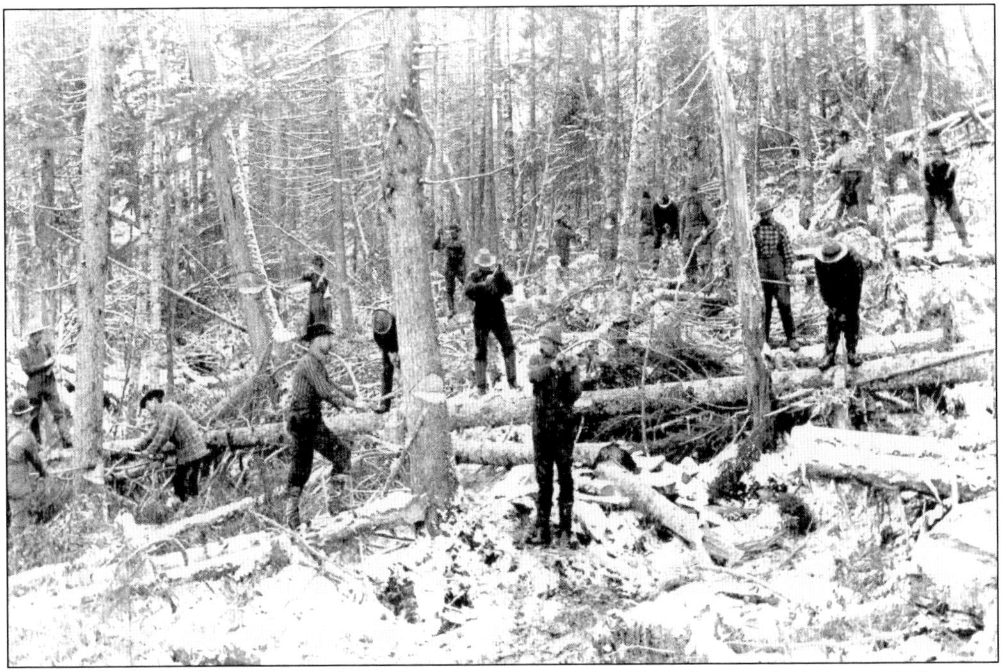

Cutting was usually done in the fall and early winter, after the sap had stopped flowing. Lumberjacks used only axes until the 1890s, when the two-man crosscut saw was introduced. With the new saws, two men could cut 100 trees a day. The photograph dates from 1888.

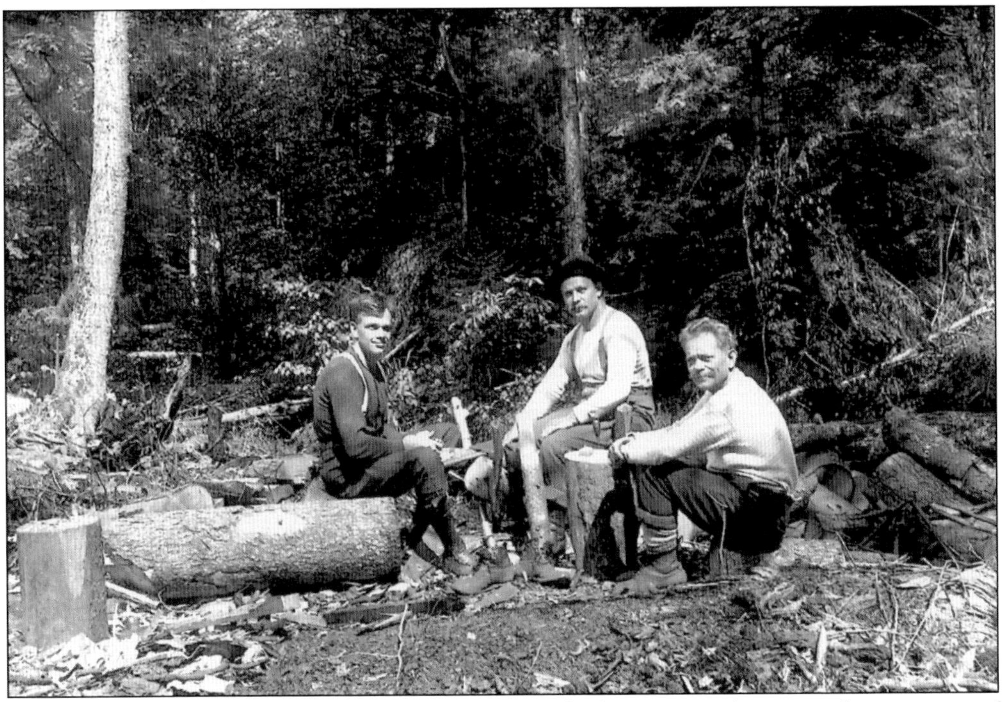

In the 1880s, an acre of Adirondack land sold for as little as 70¢ and contained an average of 16 large spruce, about all a man could cut in a day. Crosscut saws allowed loggers to cut trees closer to the ground than they could with axes.

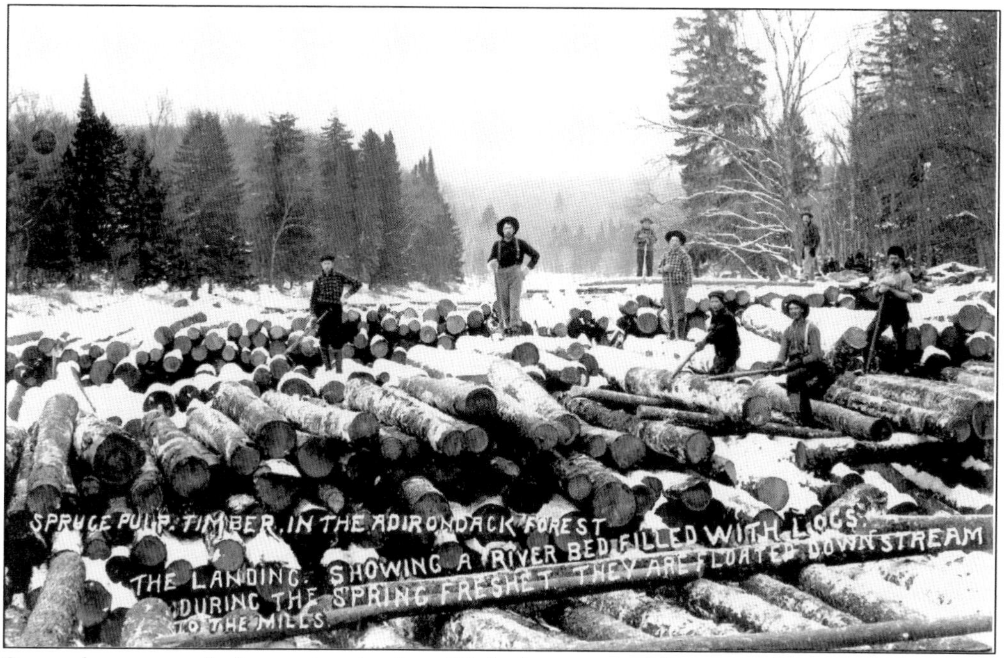


Spruce logs are stacked on a river landing. With the arrival of the spring thaw, the water would rise and the logs would float downstream to the mills. (Courtesy of George Davis.)

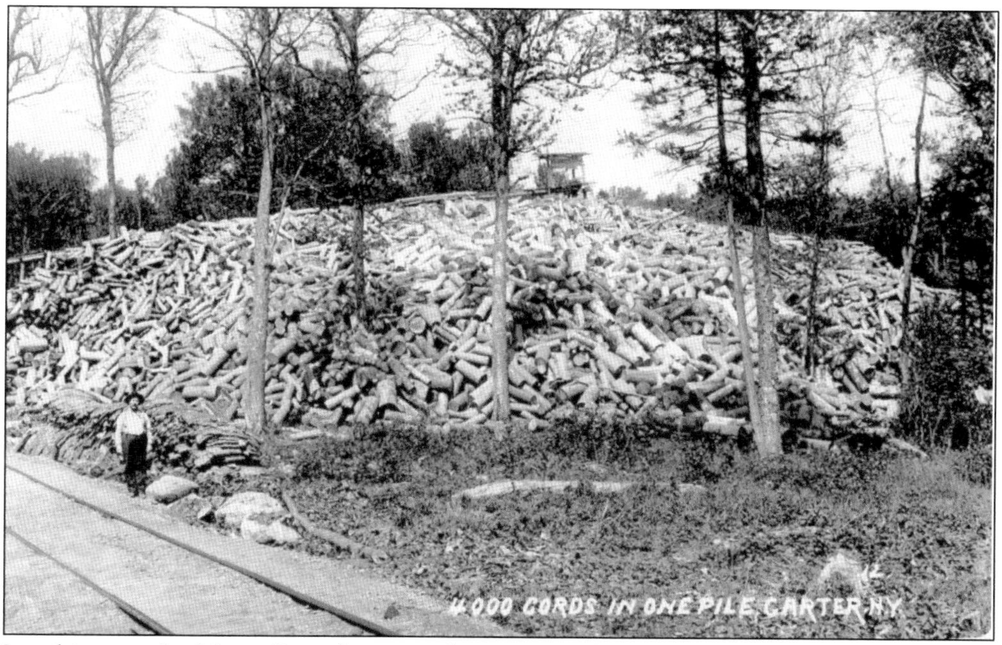


Log drives on the Moose River began in the 1860s, with the last log drive occurring in 1948. Log drives ended, in part, because paper mills preferred the shorter 4-foot pulp sticks to the traditional 13-foot saw logs. (Courtesy of George Davis.)

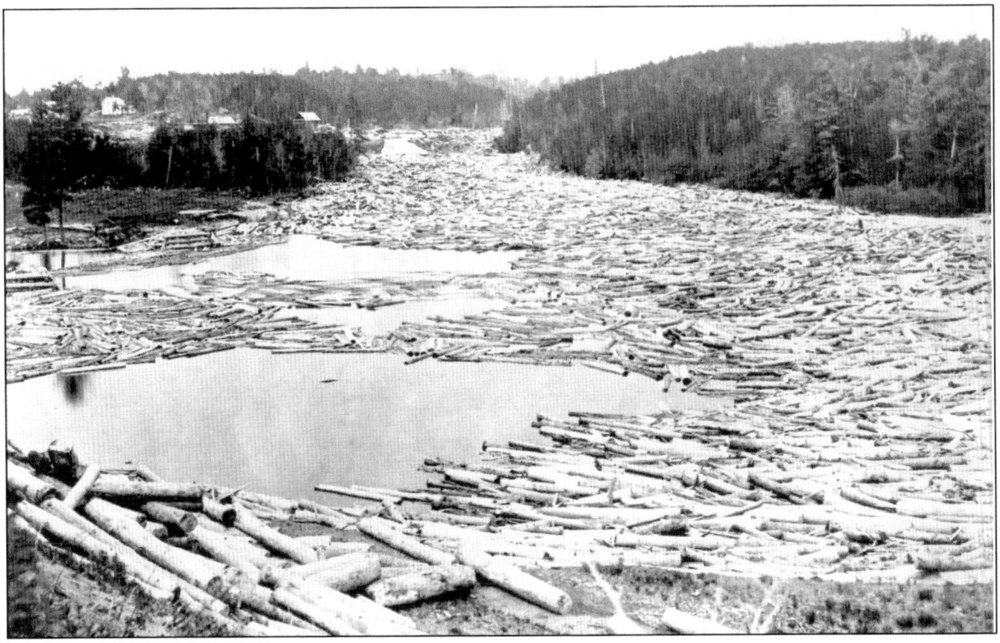

Even as woodsmen relied on the canal for transportation, canallers relied on the forest for freight. Despite their economic interdependence, woodsmen and canallers competed for the same water. When water was too low for log drives, lumberjacks broke open dams, appropriating water from the canal's vast impoundment system. The photograph dates from 1896.

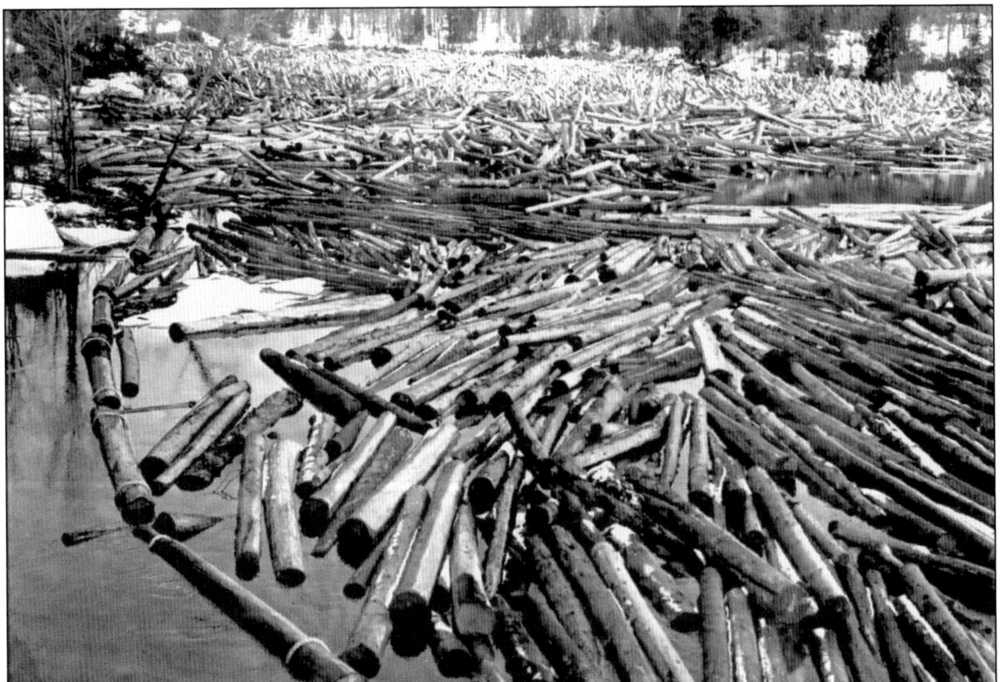

Booms on the river made it possible to collect the logs at the mill. Logs typically floated downstream at two miles per hour, but where the current was swift, they could travel 30 miles in half a day.

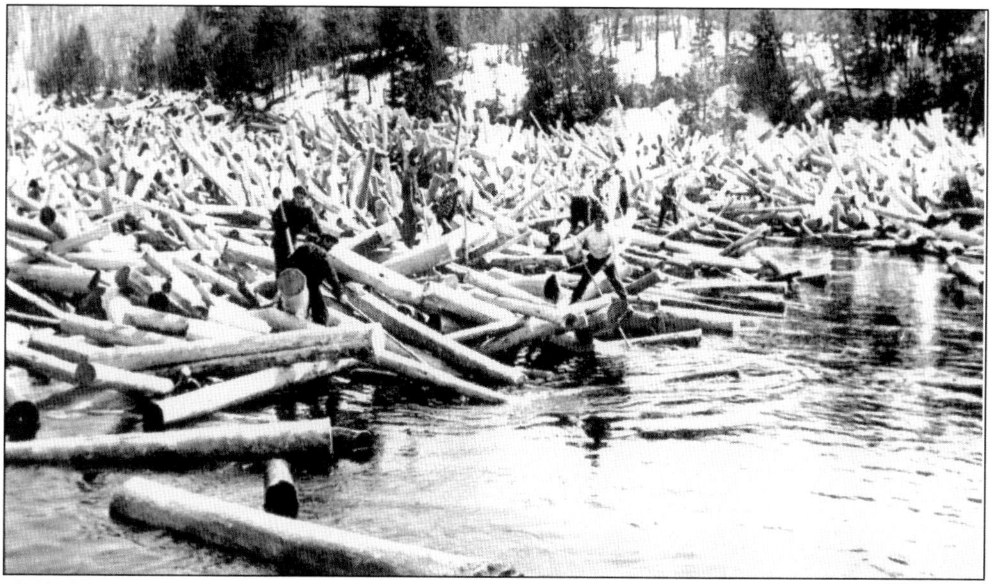

Skilled log drivers in 1890 could earn $2.50 for a day's work that lasted from 5:00 a.m. to 8:00 p.m. Drivers were equipped with a cant hook, peavey, or pike, and they wore heavy, calked shoes called croghans, named for the Lewis County village in which they were made.

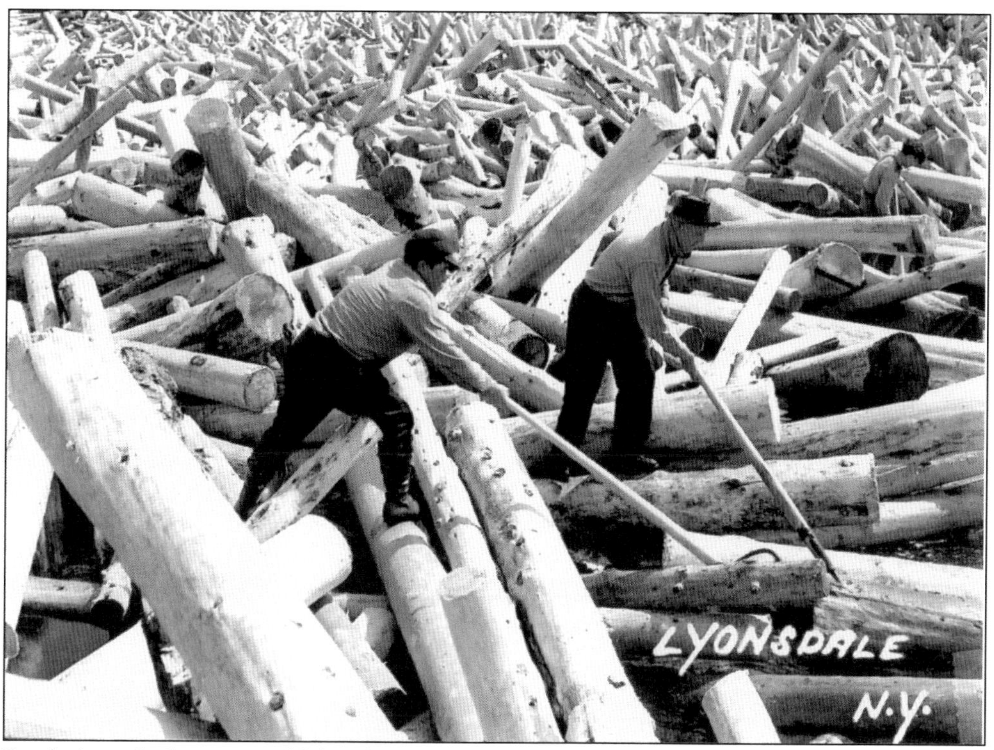


By day's end, drivers would be thoroughly soaked. To dry them out, cooks built roaring campfires next to temporary shelters on the riverbank. Rheumatism was common, and drivers older than 35 were rare.

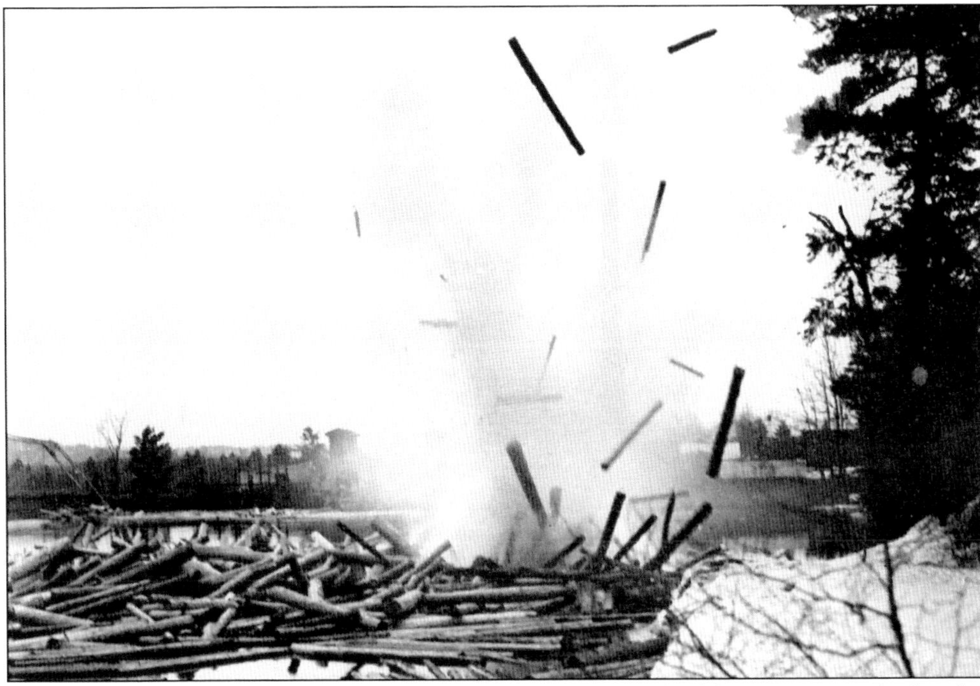

Logjams could form when a single log became snagged on a boulder. Removing the key log was extremely hazardous work, and deaths were fairly common. When drivers were unable to remove the key log manually, dynamite was used, as it is here on Moose River.

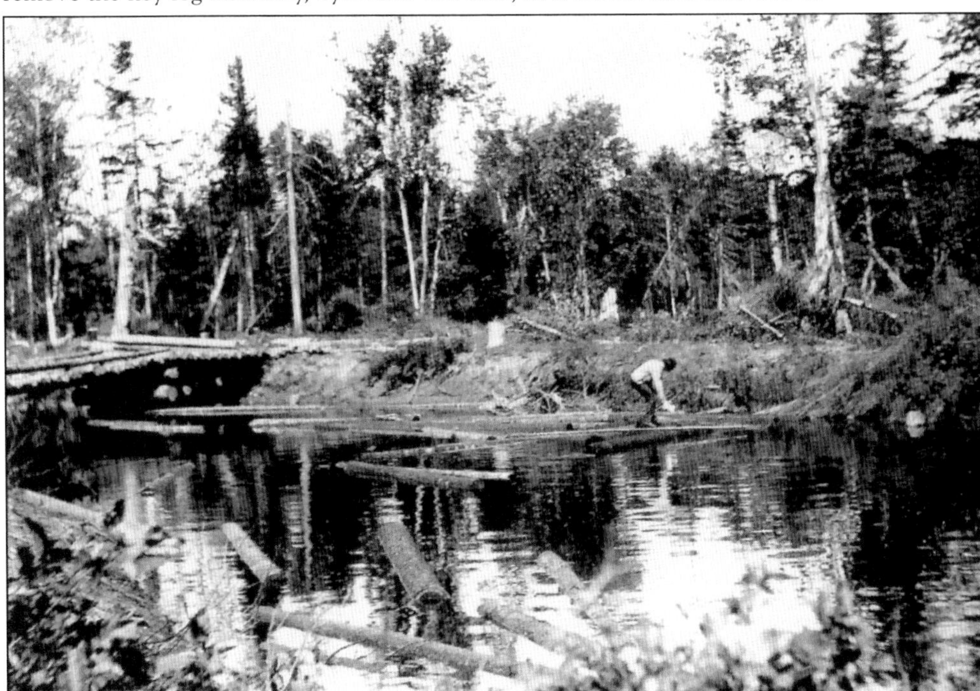

Upon completion of the main drive, loggers would conduct fly drives to collect stray and stranded logs. The scene pictured here is at Moulin. (Courtesy of Town of Webb Historical Association.)

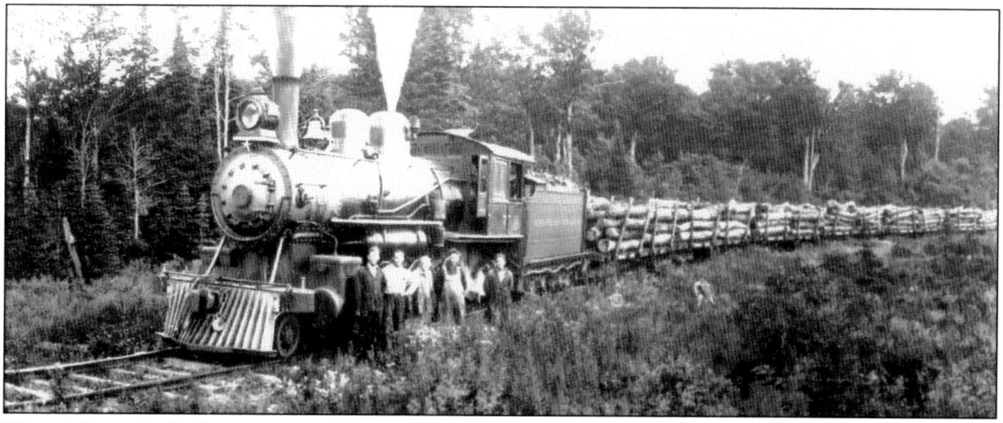

The Adirondack forests were on the eastern side of the canal, but extensive logging also occurred on Tug Hill Plateau, west of the canal. Beginning in 1902, Henry Gould hauled timber from Page on the Glenfield and Western Railroad.

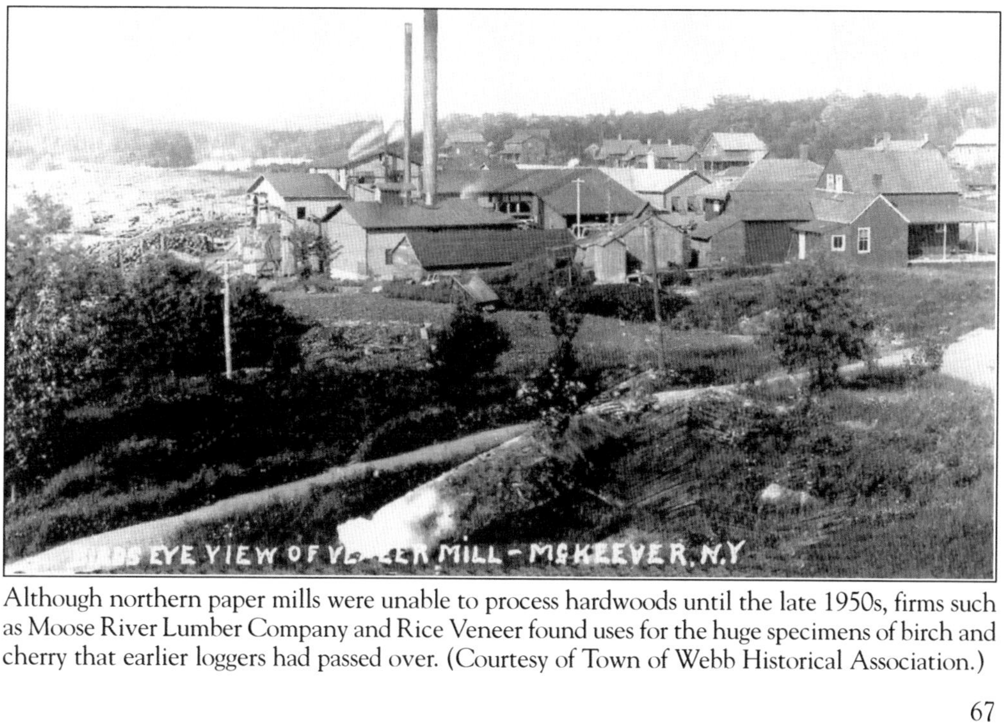


Although northern paper mills were unable to process hardwoods until the late 1950s, firms such as Moose River Lumber Company and Rice Veneer found uses for the huge specimens of birch and cherry that earlier loggers had passed over. (Courtesy of Town of Webb Historical Association.)

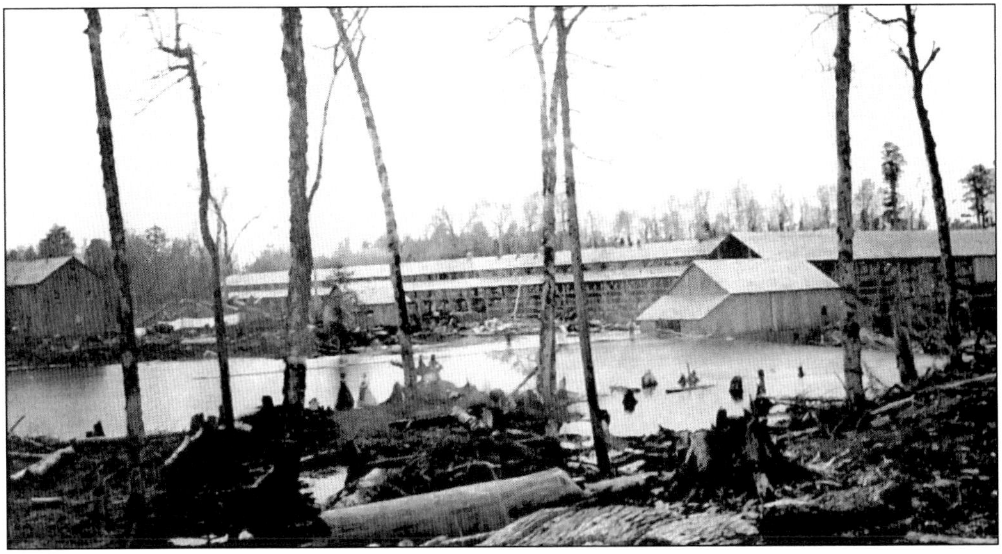

Mammoth Tannery was one of the largest tanneries in the Adirondacks. Built by Lyman Lyon and H. D. H. Snyder in 1866, this tannery was located at Moose River Settlement, north of Port Leyden. Tanneries were situated near stands of hemlock, but because they also had to be located within a day's drive of a canal or rail line, the tanneries tended to spring up on the periphery of the Adirondacks. (Courtesy of Town of Webb Historical Association.)

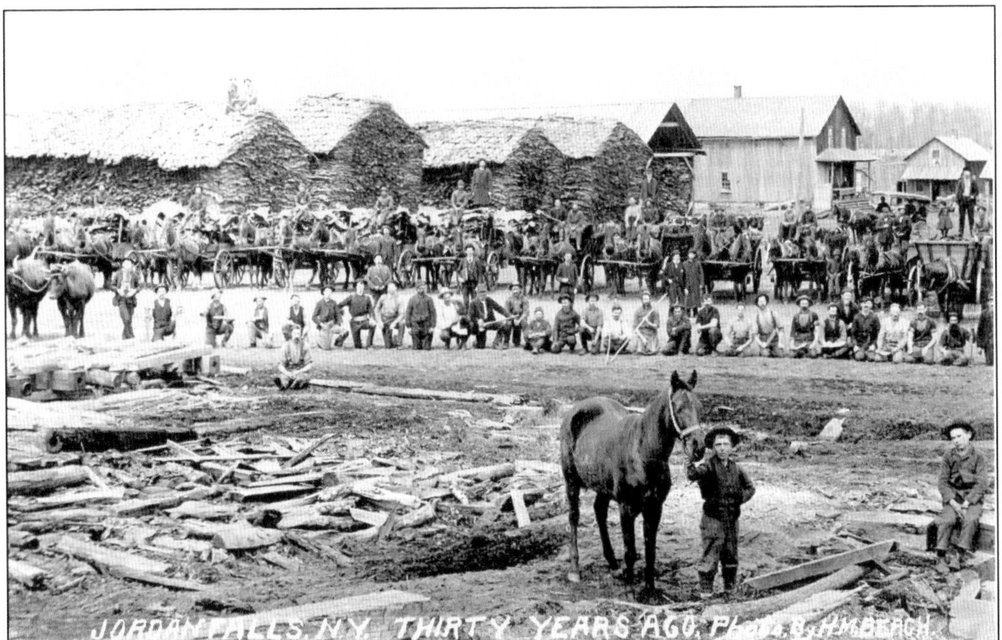


At tanneries like this one at Jerdan Falls, bark was stacked in narrow shedlike piles, often 300 feet long. An entire cord of bark was needed to tan four or five hides, and larger tanneries might consume 6,000 cords a season. To sustain production for 20 years, these tanneries might require 50,000 acres of land. An estimated two-thirds of all hemlock trees were left to rot once the bark had been removed. (Courtesy of George Davis.)

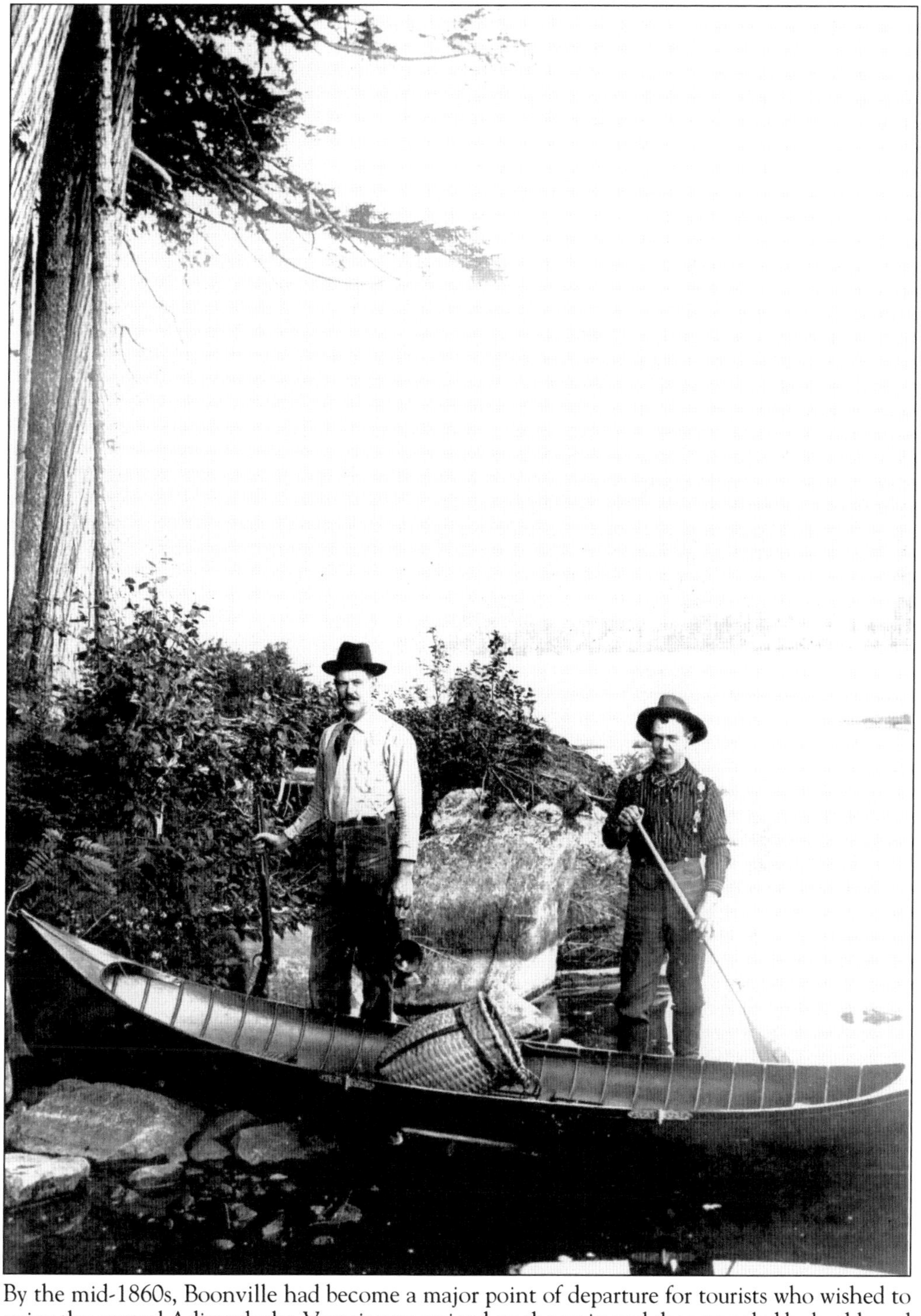

By the mid-1860s, Boonville had become a major point of departure for tourists who wished to enjoy the central Adirondacks. Vacationers arrived on the train and then traveled by buckboard along Brown's Tract Road to the Fulton Chain of Lakes. Thus, even as the canal made logging profitable, the railroad allowed tourists to witness the transformation taking place in the great northern forest. And many did not like what they saw.

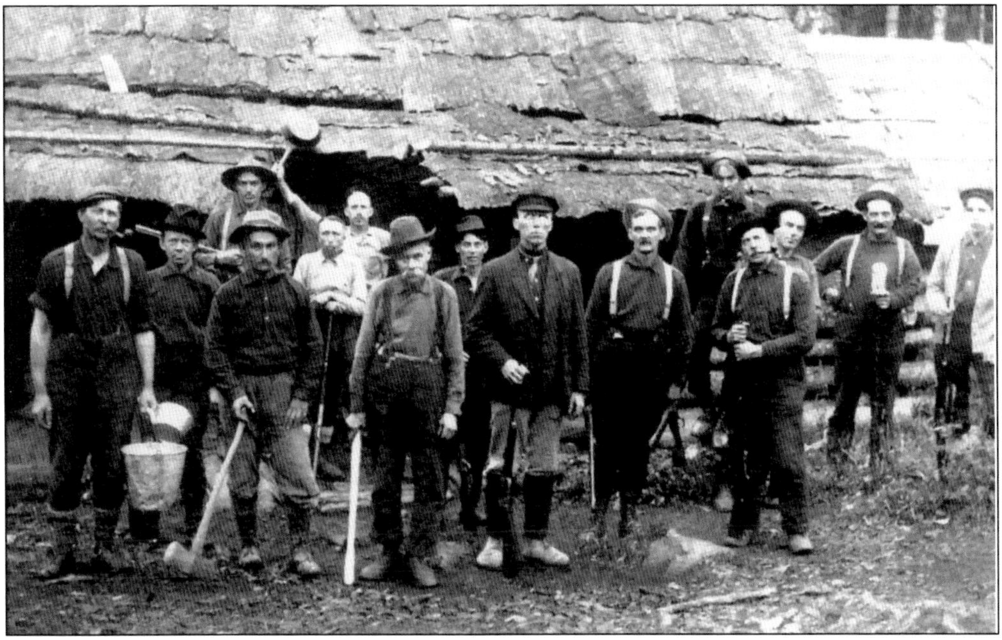

In the summer, men who enjoyed the sporting life might work as Adirondack guides. In the winter, the same men might work as loggers. (Courtesy of Town of Webb Historical Association.)

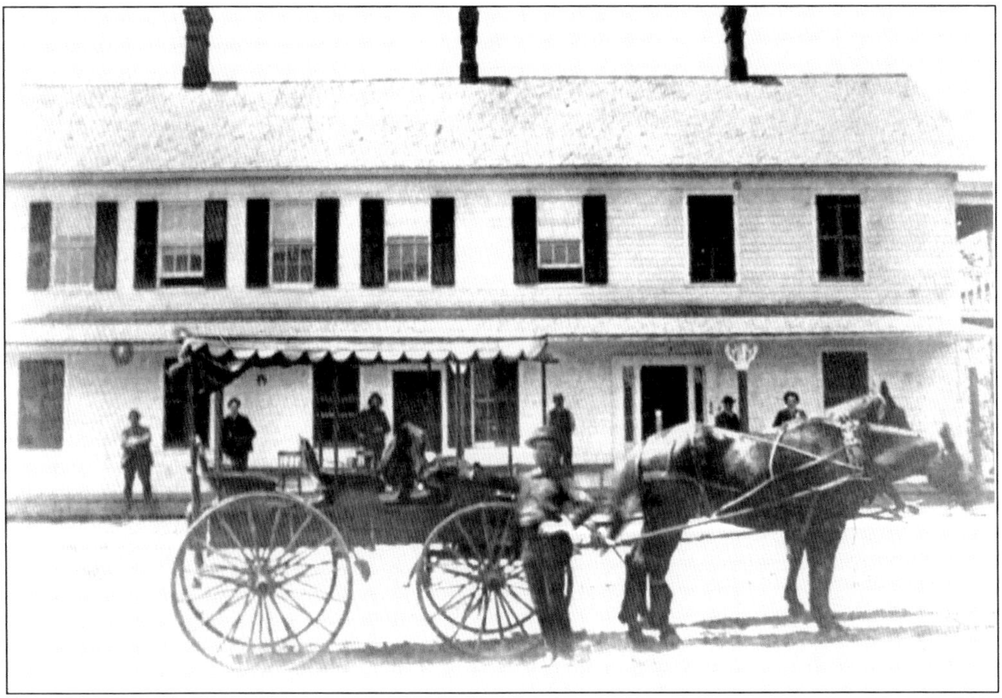

In his covered buckboard, Jim Higby (in foreground) carried tourists from Boonville over Brown's Tract Road to the Moose River Hotel. This trade largely disappeared in 1892, when William Seward Webb completed the Mohawk and Malone Railroad to Thendara, allowing vacationers to visit the central Adirondacks in relative comfort. Spur lines from the new railroad opened the region to both softwood and hardwood logging.

# *Four*

# Canal Town

The Black River Canal's economic impact on the North Country was both immediate and long-lasting. Warehouses, shipping houses, hotels, taverns, foundries, feed stores, and harness shops sprang up in villages all along the canal. Boatbuilding became an important local industry. The original boats were "foreign"—imported from other canals throughout the state—but before long, most boats were built, owned, and manned by locals. Operating the canal itself was an important source of employment. Lock tenders were hired, and large maintenance crews were employed to mend the frequent breaks in canal walls, repair lock equipment, patrol the banks, clear landslides, fix sinkholes, and remove debris that washed into the canal.

The new prosperity was evident in downtown Boonville. In the years following the Civil War, blocks of handsome, three-story brick and wood-frame buildings, many with high false fronts and elaborate ornamentation, replaced earlier commercial structures. Increased trade, spurred by the canal, led to wealth for many local merchants, and handsome residences in the Italianate Revival style, replete with cupolas, began to dominate village streets.

As local forests were cleared, farming became an increasingly important industry. Cheese and butter were shipped on the canal, as were various grains and root vegetables. Over time, potatoes became the dominant export crop, although they never rivaled the export of timber. In 1880, 152,000 bushels were shipped on boats, each carrying a standard load of 3,000 bushels.

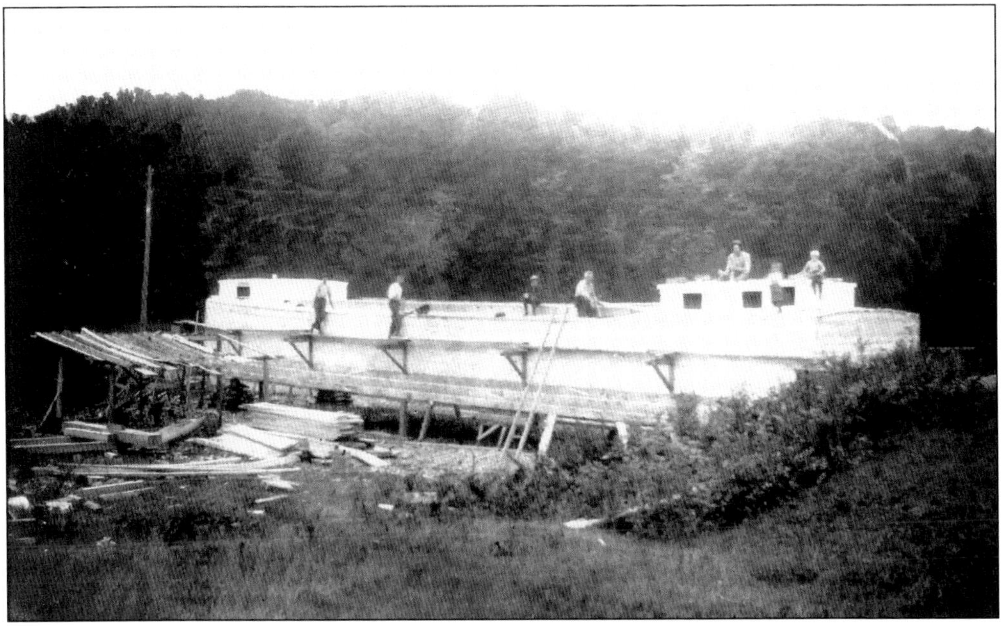

The earliest boats on the Black River Canal came from other areas of the state. But by the mid-1860s, boatbuilding had become an important local industry. Small firms, like those in Boonville, could turn out two or three boats a year. In the 1870s, boats sold for between $1,100 and $1,700. (Courtesy of Carthage Free Library, Heritage Room.)

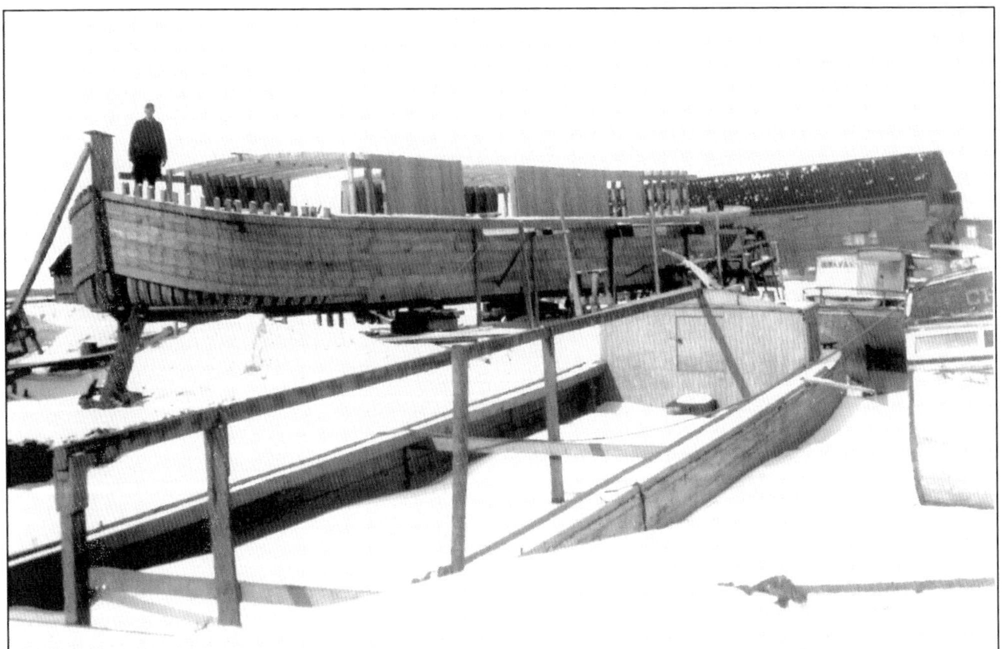

The boat under construction is probably a bullhead. The bullhead deck was covered by a cabinlike structure that protected perishable goods such as grain and flour. In Boonville, the first boatbuilder was Sam Ferguson, who established his dry dock on Canal Street near Lock 71 about 1858. The photograph dates from the 1890s.

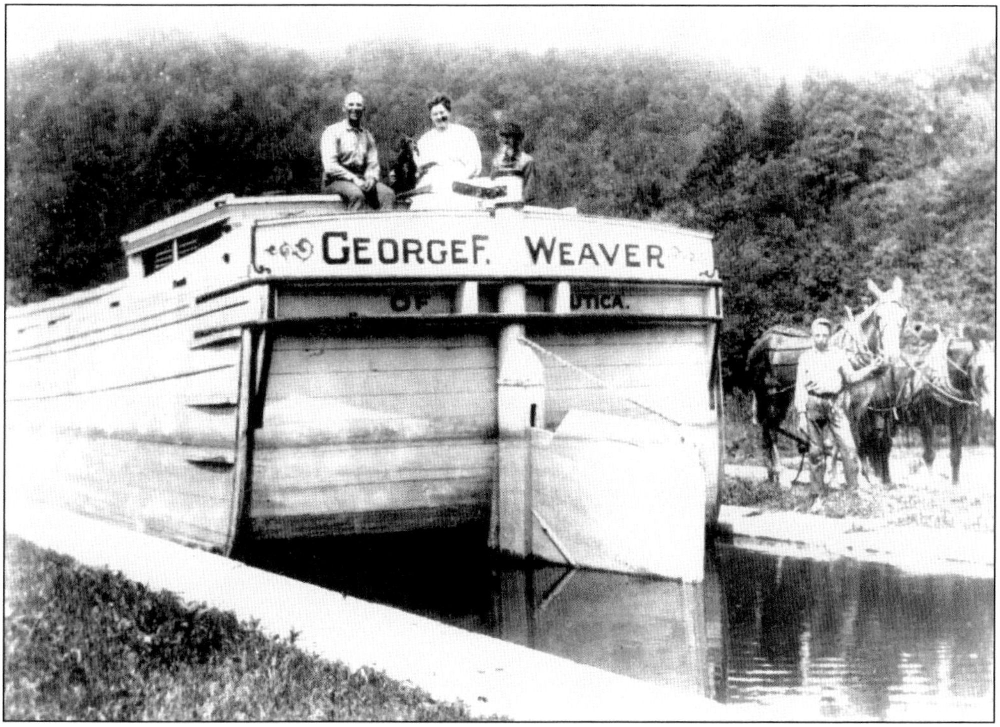


George Seiter built the *George Weaver* for William Richardson of Utica. Seiter's shop was on Mill Creek just north of Lock 71 in Boonville. His three brothers also built boats, most notably Frank Seiter, who claimed to have built 40 boats during his career. The bearded man aboard (right) is Harvey Nestle, who tended Locks 37 and 38 during Republican administrations.

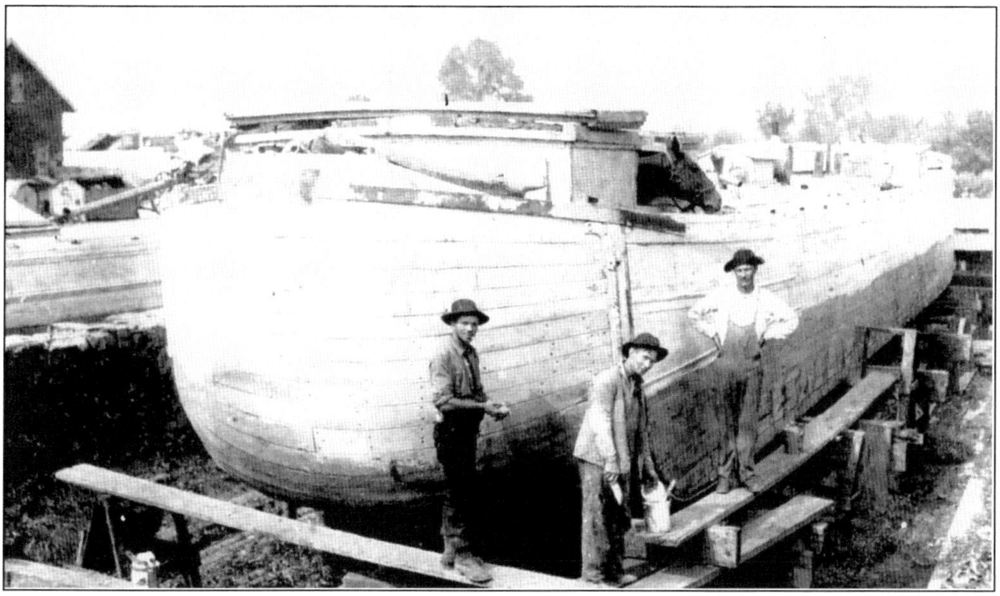

With regular upkeep, a canal boat typically lasted from 20 to 25 years. Boatbuilders typically used white oak for the bow, pine for the sides, and hemlock for the inside timbers. Construction and repair work was done in the open air. Notice the mule aboard the boat.

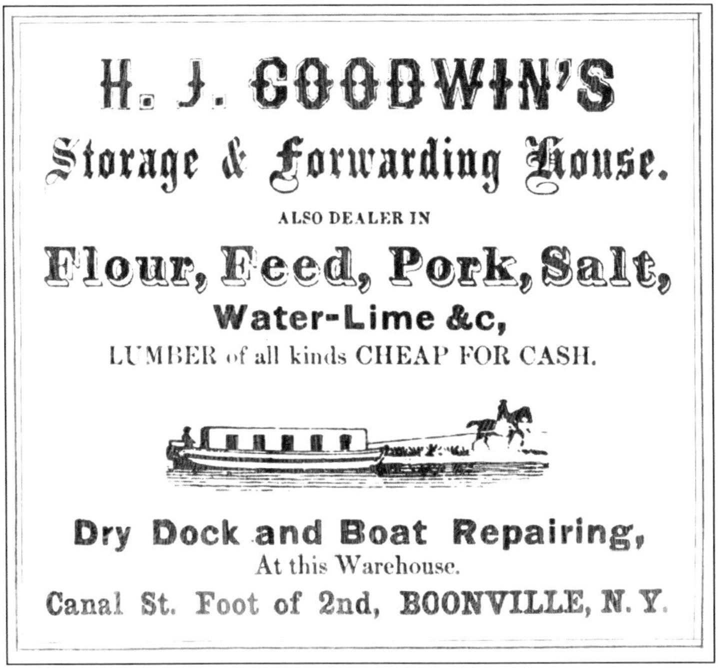


H. J. Goodwin built and repaired boats at the site of the former Joslin and Schultz warehouse on Canal Street.

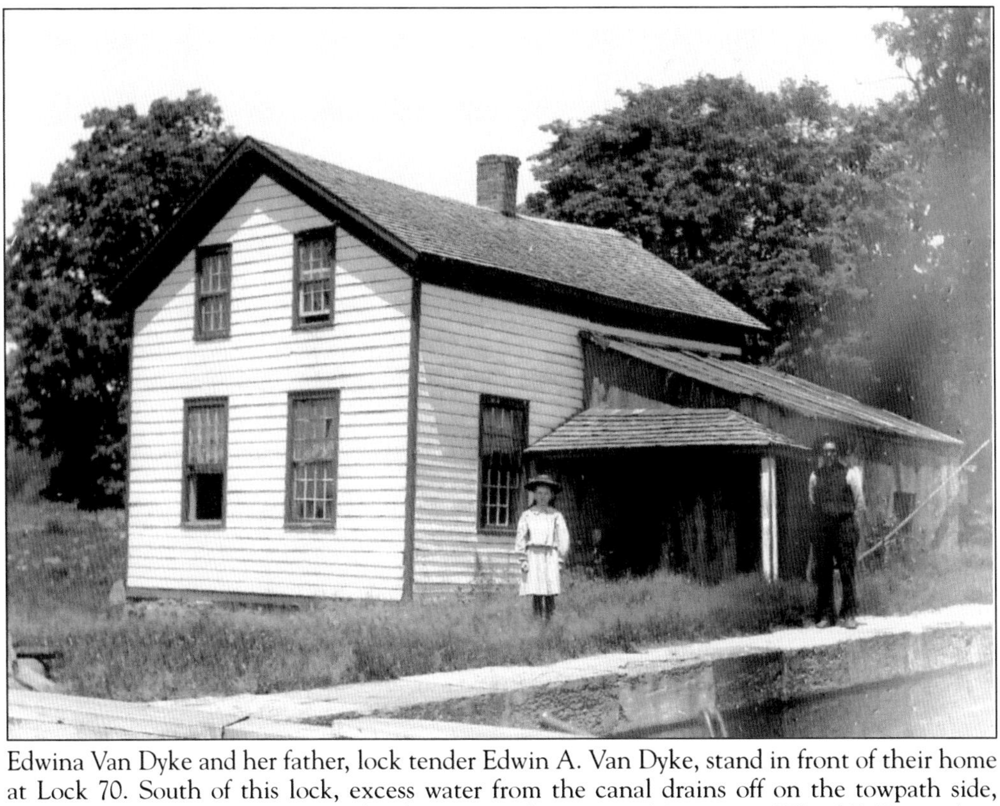

Edwina Van Dyke and her father, lock tender Edwin A. Van Dyke, stand in front of their home at Lock 70. South of this lock, excess water from the canal drains off on the towpath side, forming Baker's Falls, once a popular destination for picnics. (Courtesy of Floyd E. Draper.)

Opposite Baker's Falls (pictured here), the Lansing Kill forms a second waterfall as it cascades over a high limestone ledge. Waters from the two falls merge at the bottom of a glen to create a sinkhole pond. The photograph was taken in the 1960s. (Courtesy of New York State Museum.)

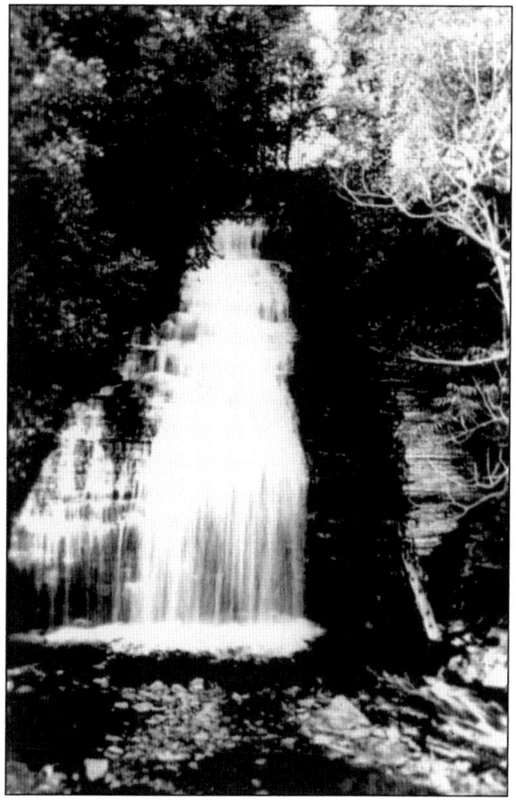

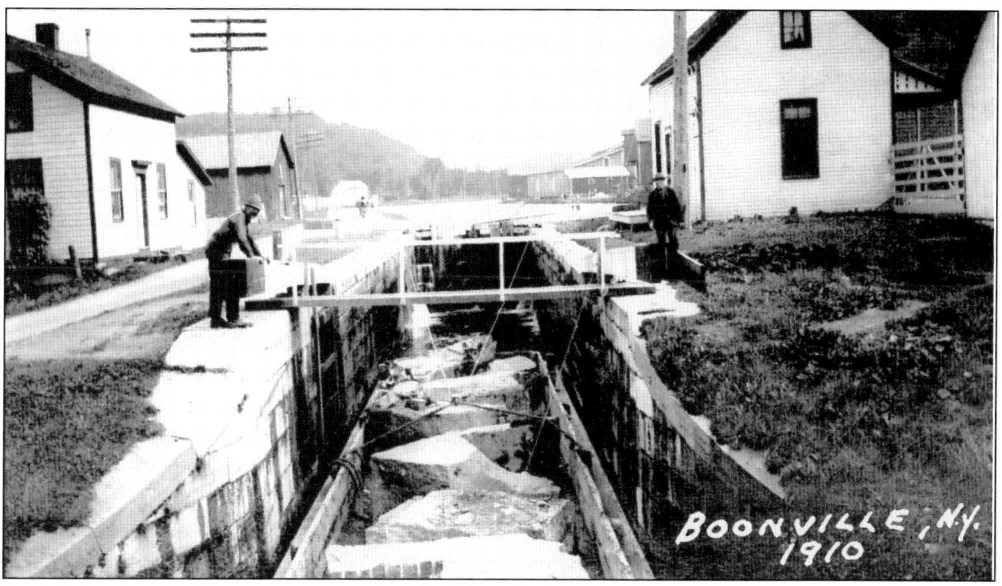


Lock 71 in Boonville sits at summit level of the Black River Canal. This lock, located just south of East Schuyler Street, was the northern terminus of the canal in 1850. Apart from a state historical marker, all traces of the lock have disappeared.

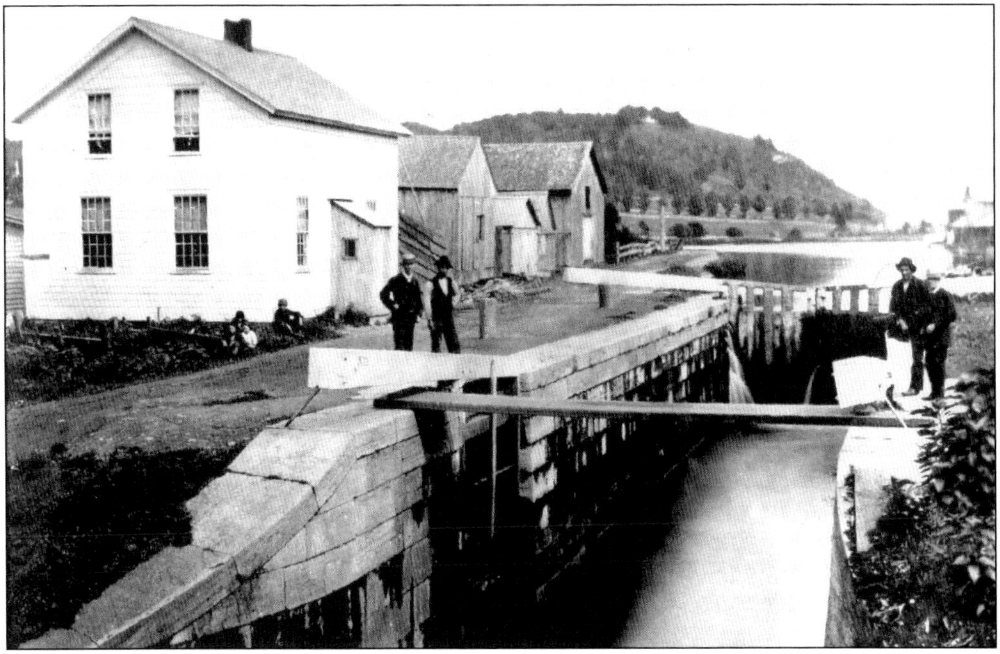

After 1882, tolls were no longer collected on New York canals. Prior to that, tolls were collected at Lyons Falls, Boonville, and Rome. The toll schedule was complex but was based on three factors: type of cargo, amount of cargo, and distance traveled. In 1867, for example, hides were assessed at 6 mills per 1,000 pounds per mile, and hemlock bark shipped at 5 mills per cord per mile.

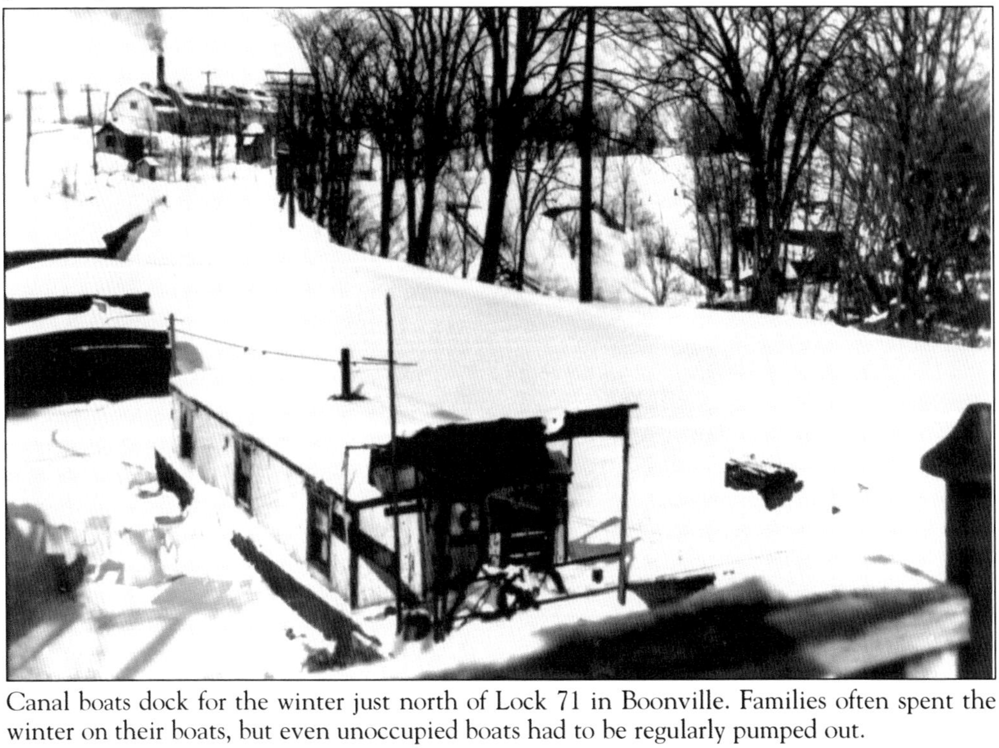

Canal boats dock for the winter just north of Lock 71 in Boonville. Families often spent the winter on their boats, but even unoccupied boats had to be regularly pumped out.

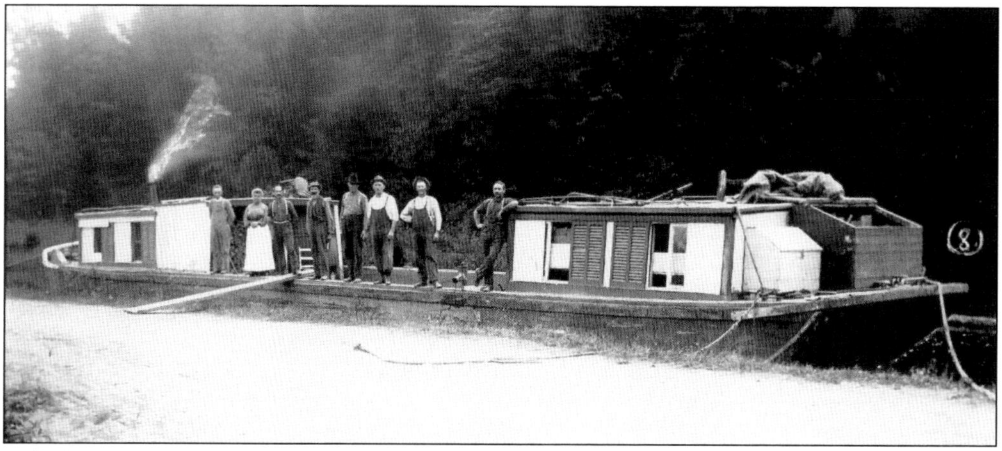

Pulled by horses rather than mules, state scows, or "hurry-up" boats, were essentially empty wooden shells capable of transporting construction materials. Each scow had a cook (here, second from left). (Courtesy of Bob Hindman.)

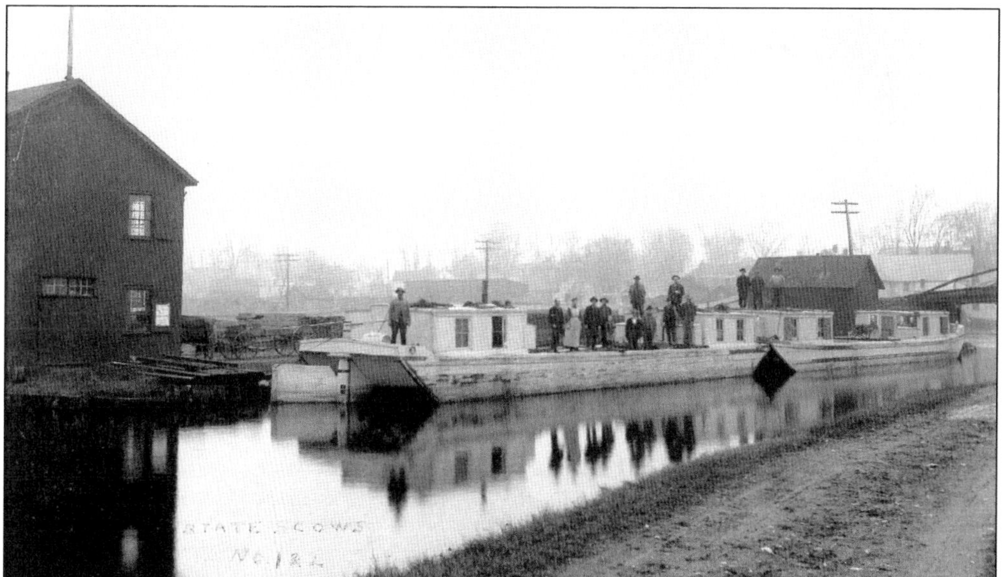


Repair crews were appointed on the basis of their political affiliation. A broadside on the wall of the state shop announces an upcoming "Grand Republican Rally." Portraits of Charles Evans Hughes and William Howard Taft adorn the windows of the far cabin. (Todd Weseloh Collection.)

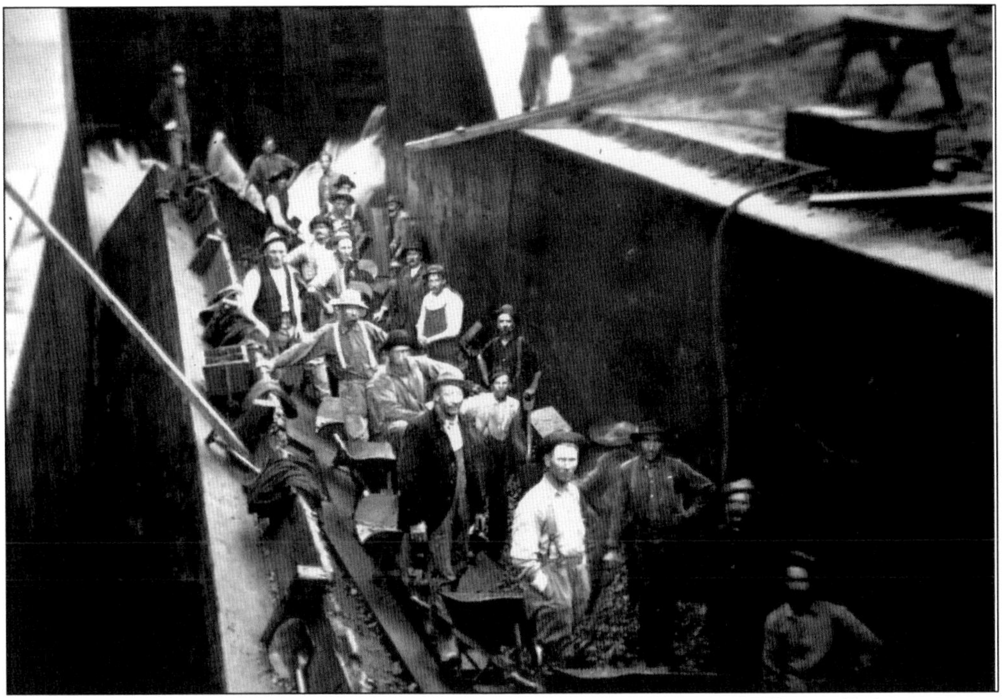

State workers unload a tipped scow inside a lock at Delta. Crews were often called on to raise two or three sunken boats a month. (Courtesy of New York State Museum.)

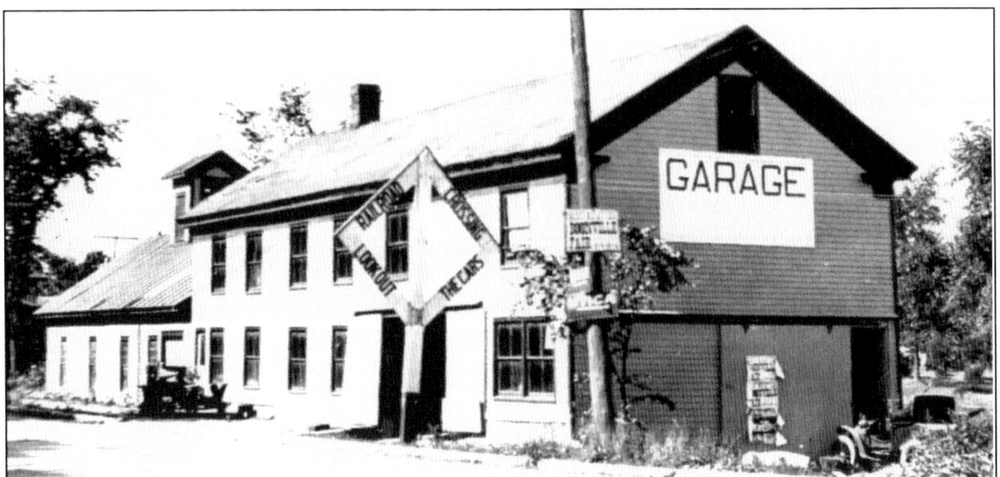


Titus Powers built a foundry at the southern end of Main Street shortly before the canal arrived. Within a few years, the foundry passed into the hands of Holdridge and Gilbert, who eventually specialized in the manufacture of farm plows. This location is the site of the future Black River Canal Museum. The photograph is from the 1920s.

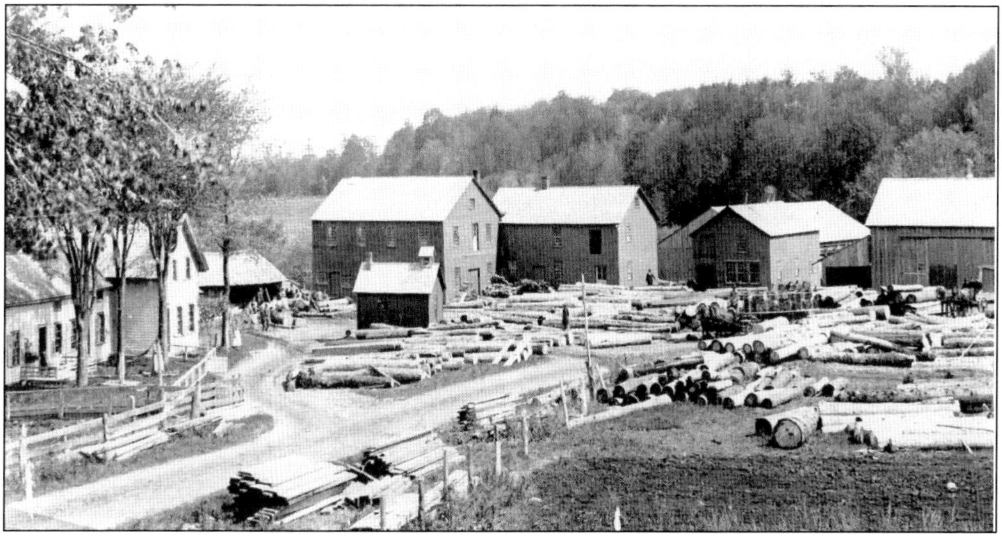

Nathaniel Sargent bought the remnants of Whiting Tuttle's sawmill on Mill Creek. In 1849, a break in the Erie Canal Feeder had washed away Tuttle's sawmill, but Sargent rebuilt the plant in 1851 and turned it into a successful chair factory. Ephraim Owens built the first sawmill on this site in 1817.

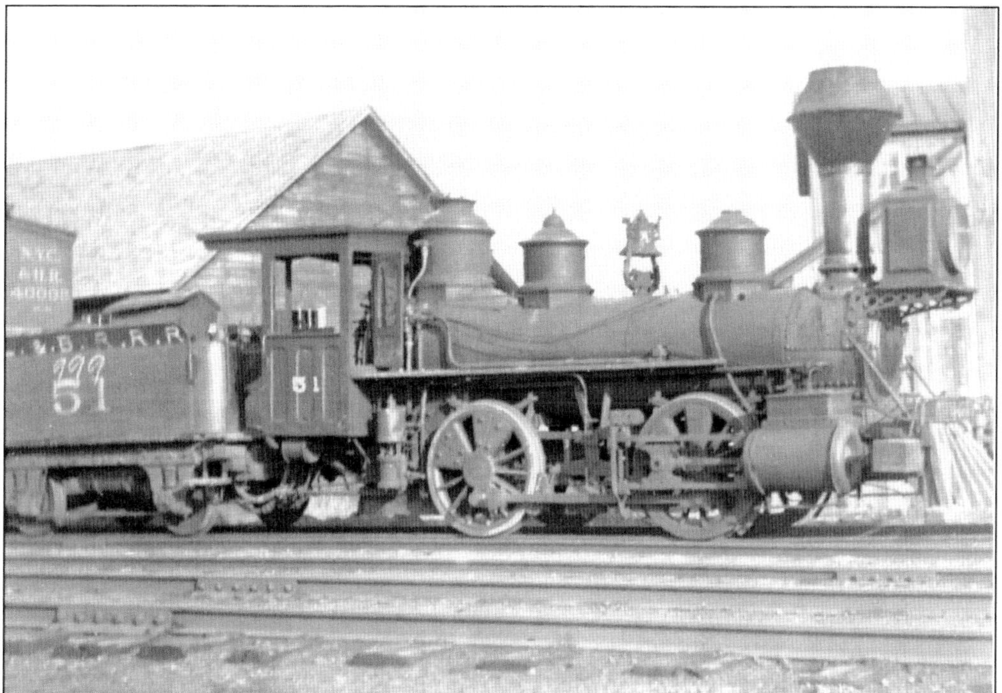


The Black River and Utica Railroad arrived in Boonville on December 13, 1855, five years after the southern section of the canal had opened.

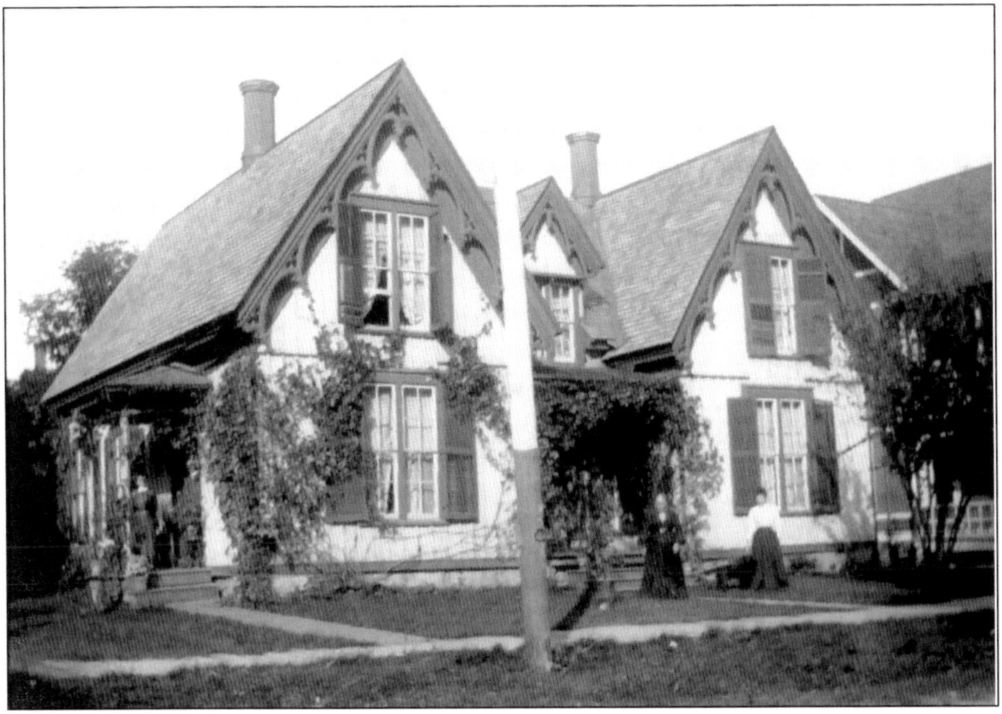

Until 1867, Boonville was the northern terminus of the railroad. The depot was at the corner of Schuyler and Post Streets, across from what is now the village park. This gingerbread house was originally the restaurant at the train station.

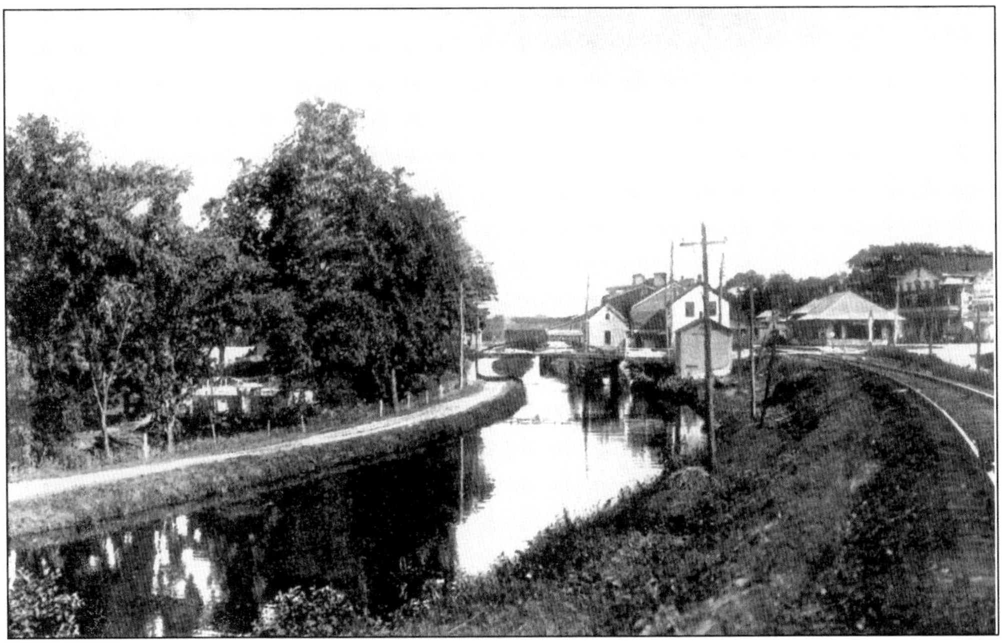

The rail line was extended north to Lyons Falls in 1867 and to Watertown in 1872. Tracks were relocated along the canal in Boonville, and a new station was built on the site of the village cemetery. Bodies were exhumed and moved to a new cemetery on Park Hill.

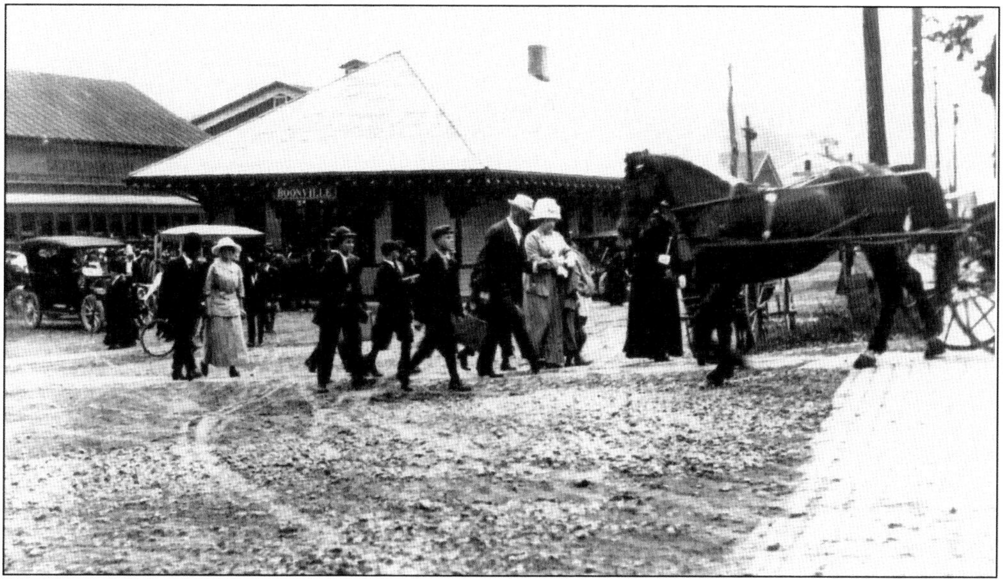


Packet boats carried passengers in the early 1850s, but the train quickly became the preferred mode of transportation. The railroad carried 78,000 passengers its first year, with four trains running daily between Boonville and Utica. The canal, however, continued to dominate the freight business. On the right are Garry and Julia Willard.

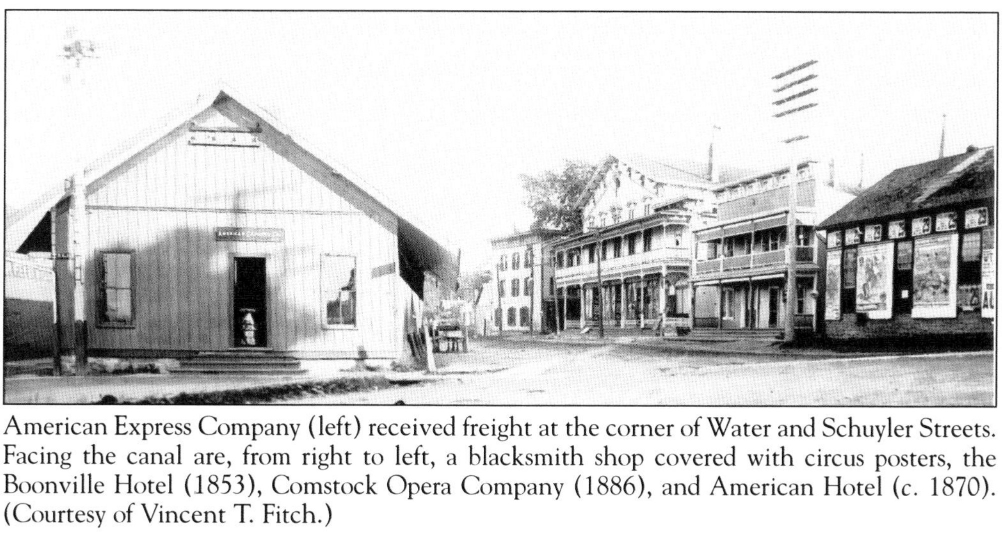

American Express Company (left) received freight at the corner of Water and Schuyler Streets. Facing the canal are, from right to left, a blacksmith shop covered with circus posters, the Boonville Hotel (1853), Comstock Opera Company (1886), and American Hotel (*c.* 1870). (Courtesy of Vincent T. Fitch.)

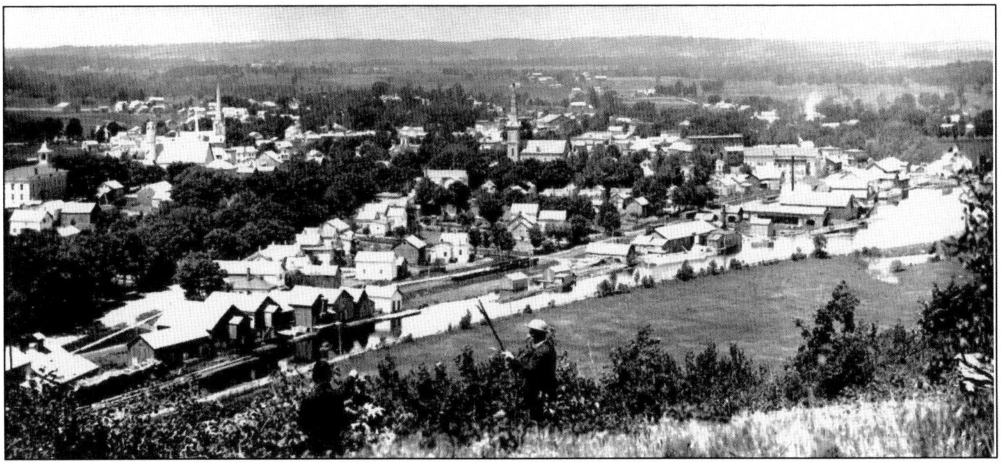

Commercial establishments sprouted along Boonville's waterfront. Besides various warehouses, grocery stores, coal yards, and dry docks, there was Thompson's steam-powered flour mill; the Black River Marble Works; and, on the east bank, Norton & Avery's door, sash, and blind factory.

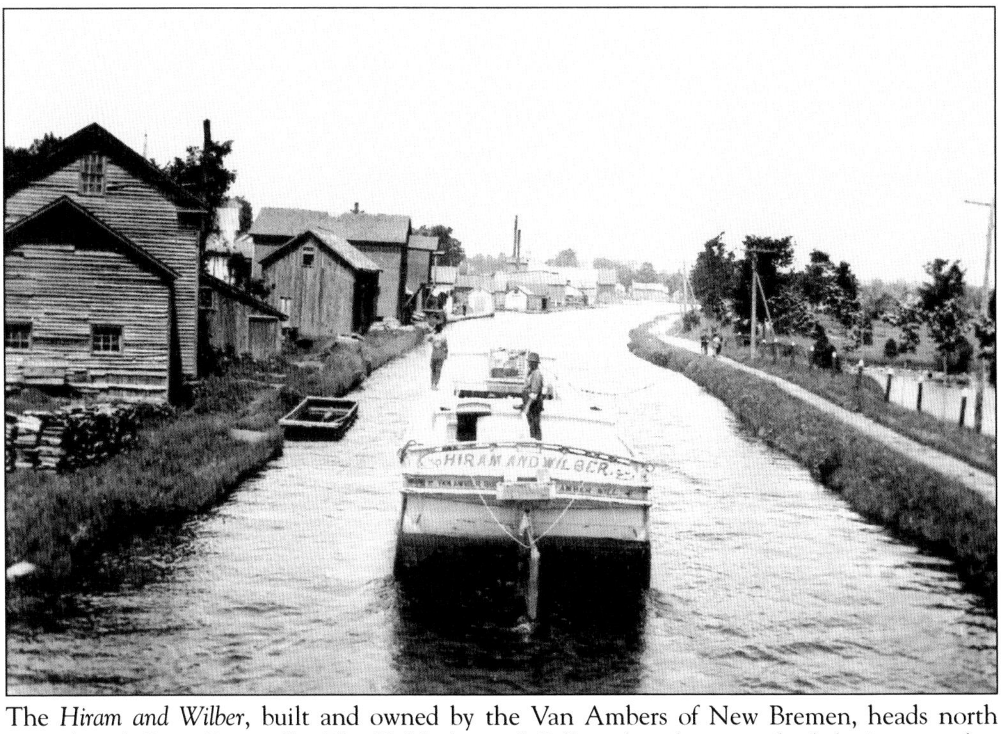


The *Hiram and Wilber*, built and owned by the Van Ambers of New Bremen, heads north toward Lock 71 in Boonville. The Holdridge and Gilbert foundry is on the left. At an earlier date, Sam Joslin and Bush Schultz's forwarding house was north of the foundry and north of that was Orsemus Clark's store. (Courtesy of Floyd E. Draper.)

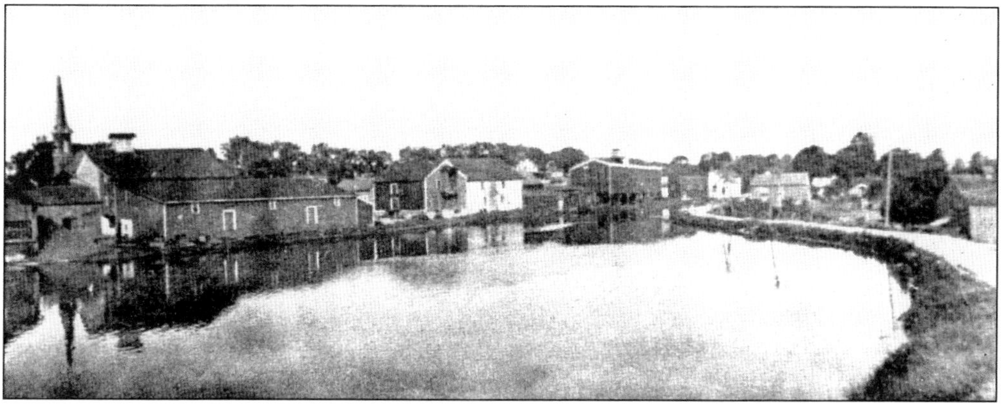

The canal widens near East Schuyler Street, providing a small harbor where boats could turn around or moor. (Courtesy of New York State Museum.)

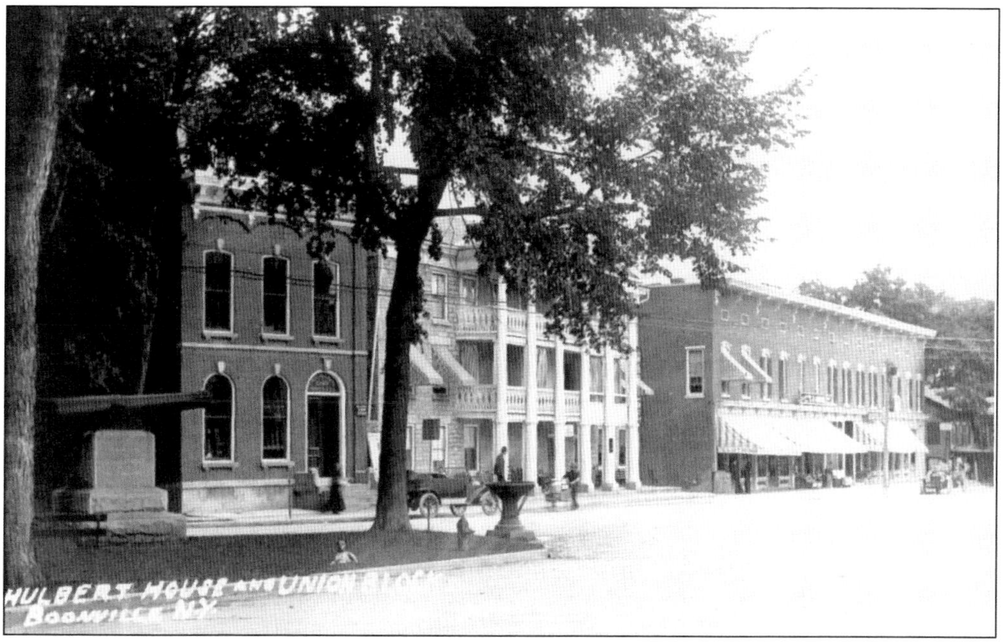


The Black River Canal brought a wave of prosperity to Boonville, and new buildings began to appear. The Bank of Boonville, on the left with the mansard roof, was erected around 1866. In the middle, the Hulbert House, with its classical portico, is an earlier structure built by Ephraim Owen about 1819. On the right is the Union Block, which was built about 1862. (Courtesy of Joann Sattler.)

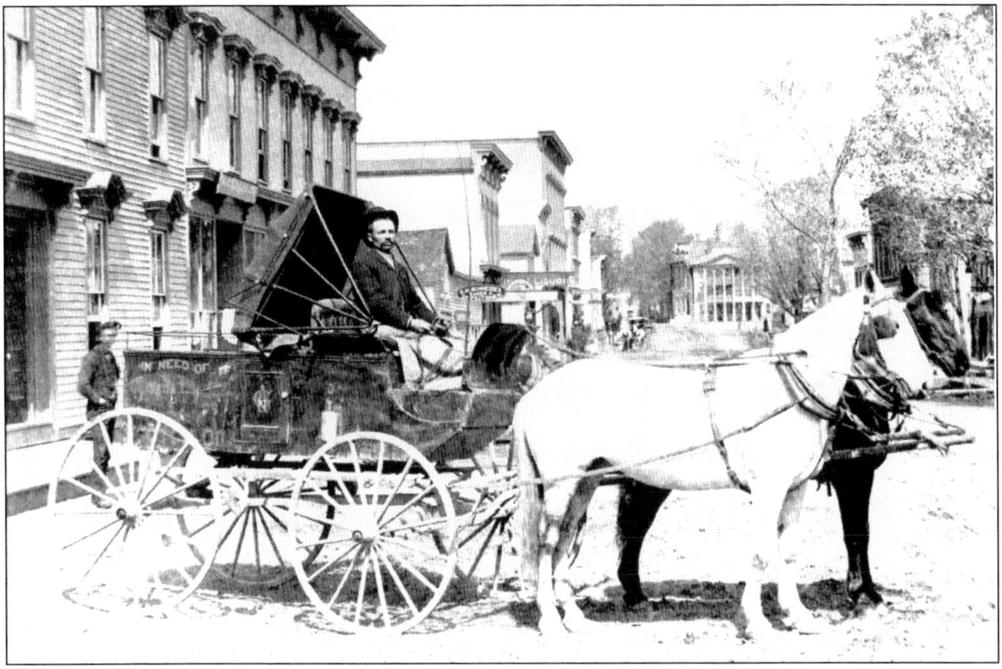

A Scotch Oil salesman from Forestport poses in his carriage on Main Street. He is parked outside the Central Hotel, built by Matthew Seiter in the 1890s. At the height of the canal era, earlier commercial structures gave way to taller buildings with horizontal cornices and ornamental brackets.

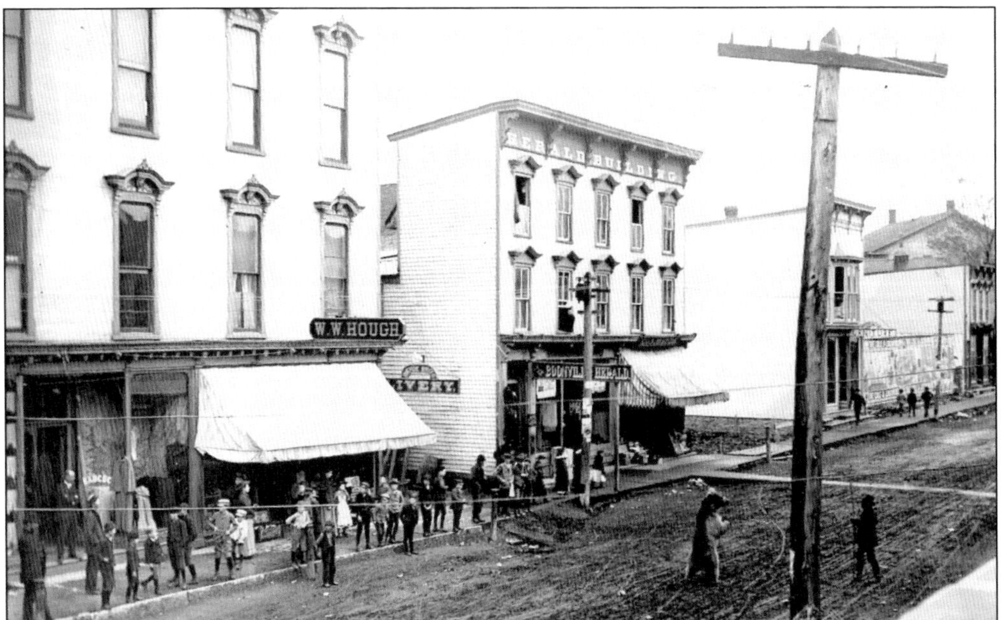


Children watch a bear perform in front of W. W. Hough's store on Main Street. Besides his store, W. W. Hough ran a kindling wood factory south of town, importing wood from Forestport on the feeder and then processing that wood for shipment. Business prospered until coal gained in popularity.

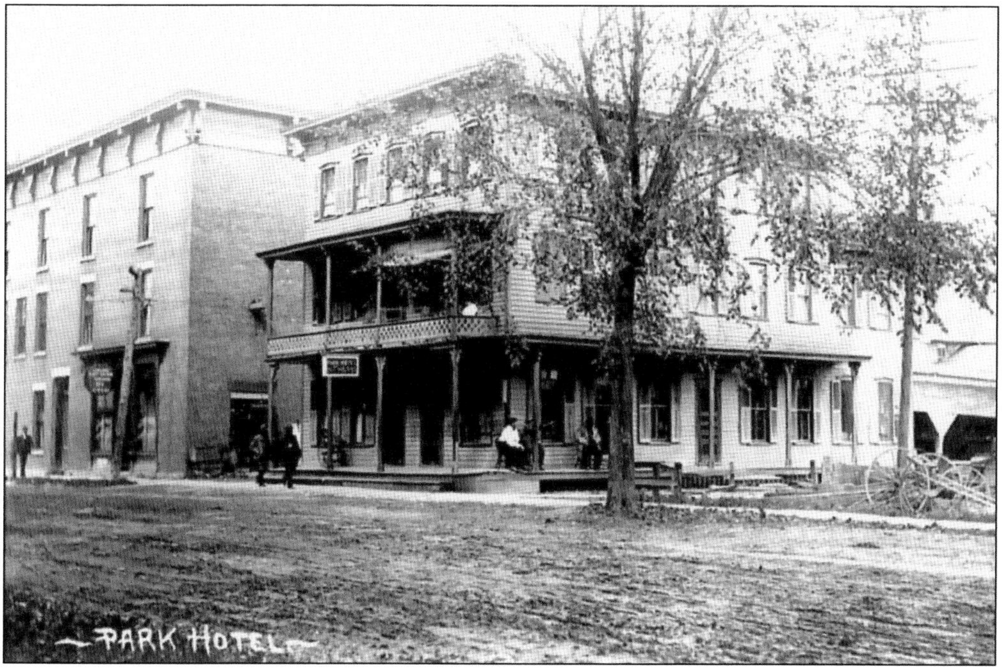


Built by George Marcy about 1890, the Park Hotel on Post Street was one of several hotels that flourished during the canal era.

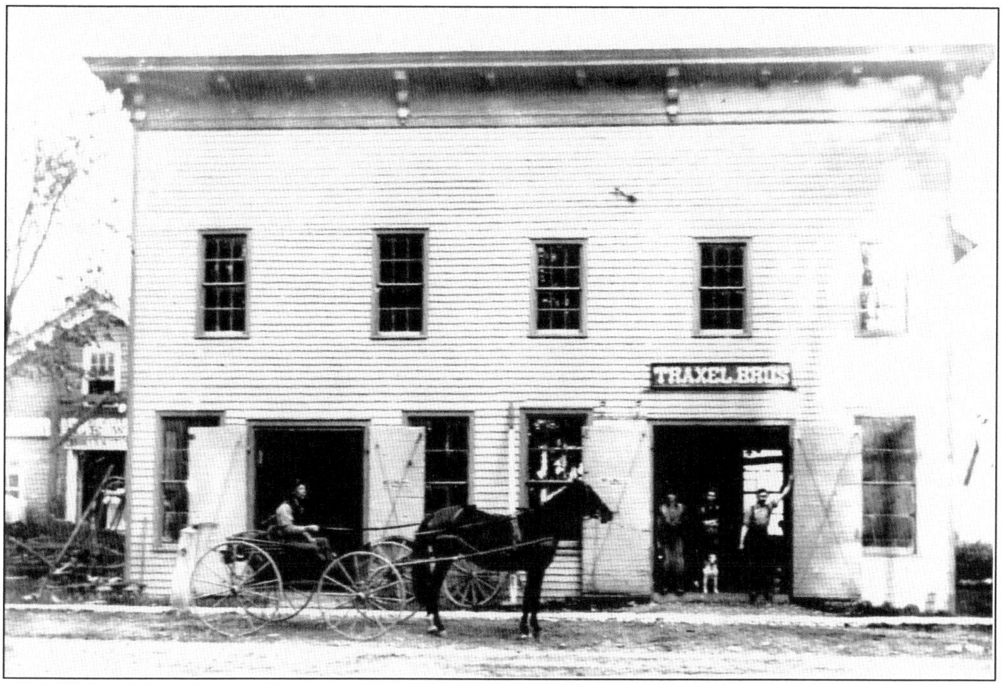


Canallers and the general population patronized Traxel's blacksmith shop. Henry Traxel and his sons George and Fred stand in the doorway. Dan Gookins sits in the buggy. The livery was located at the corner of Post Street and Railroad Avenue. The building on the left is the Mayle Granite and Marble Works.

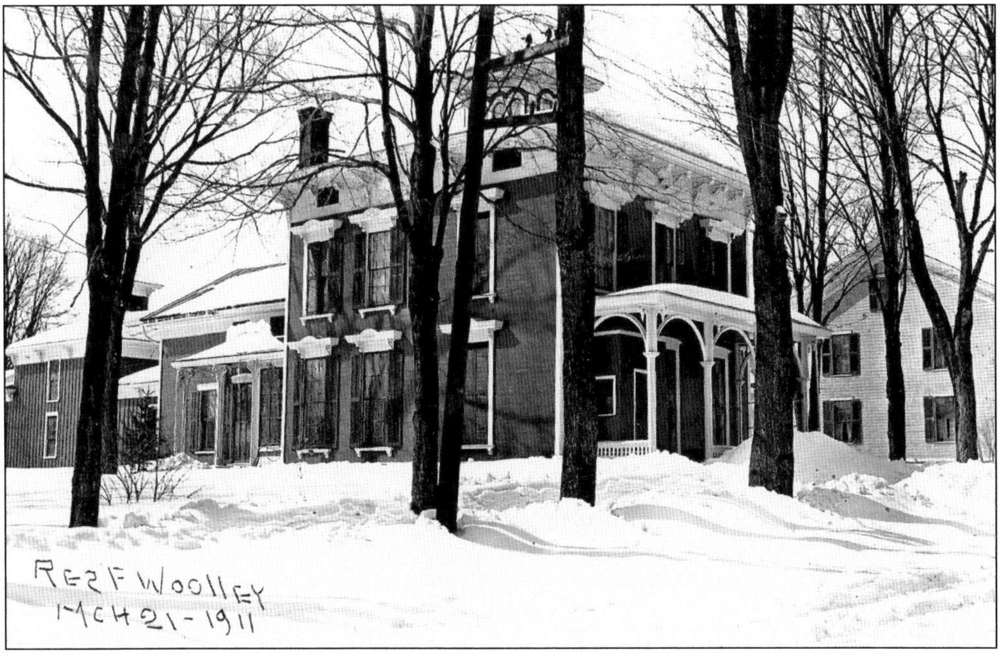


Frederick Woolley was superintendent of the Black River Canal between 1915 and 1918. His residence, with its fashionable Victorian cupola, was on Schuyler Street.

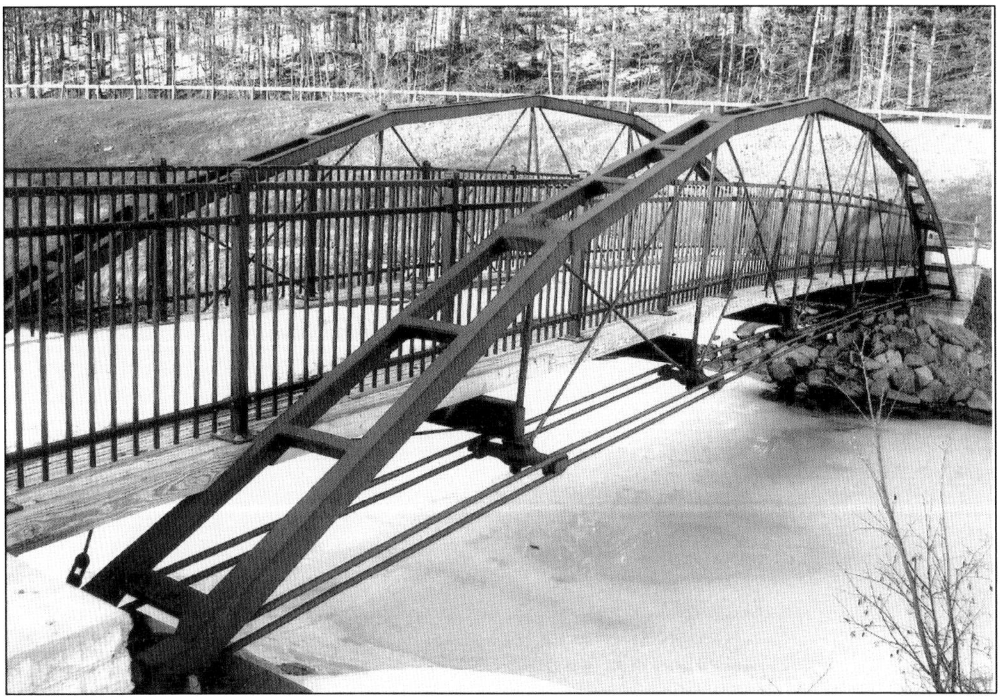

Area resident Squire Whipple patented the design for the bowstring iron-truss bridge in 1841. A "bowstring" across the bottom produces a tension that eliminates the need for buttresses or strong piers at the ends of the bridge. Whipple's design became the standard for bridges in the state canal system, but only a handful of his bridges, like this one in Boonville, have survived.

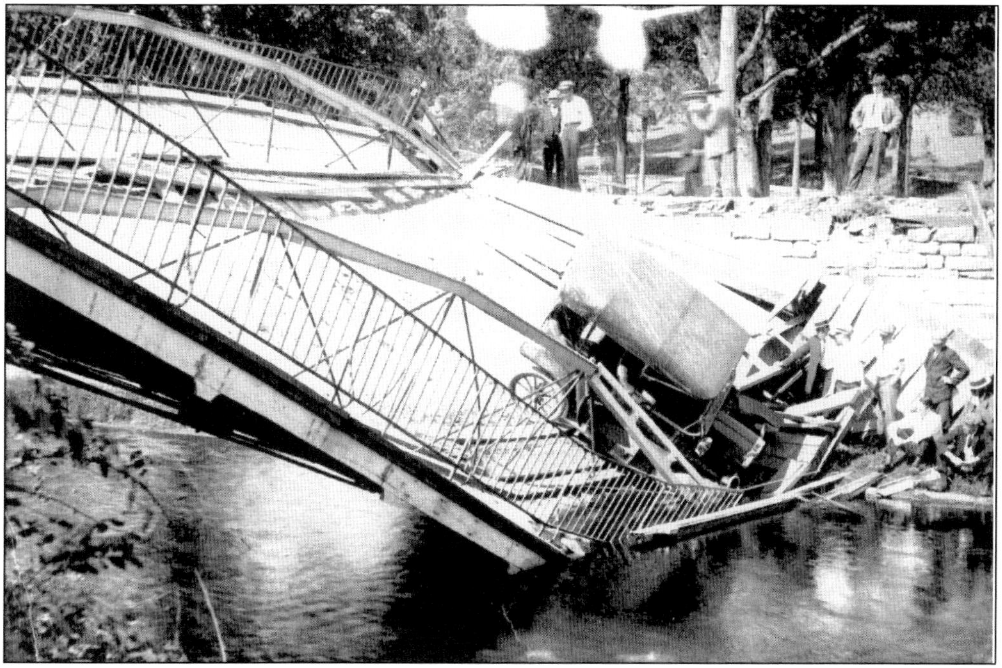

The Whipple bridge on Main Street crumples after being struck by a load of crushed stone in 1913. The Model T just happened to be in the wrong place at the time.

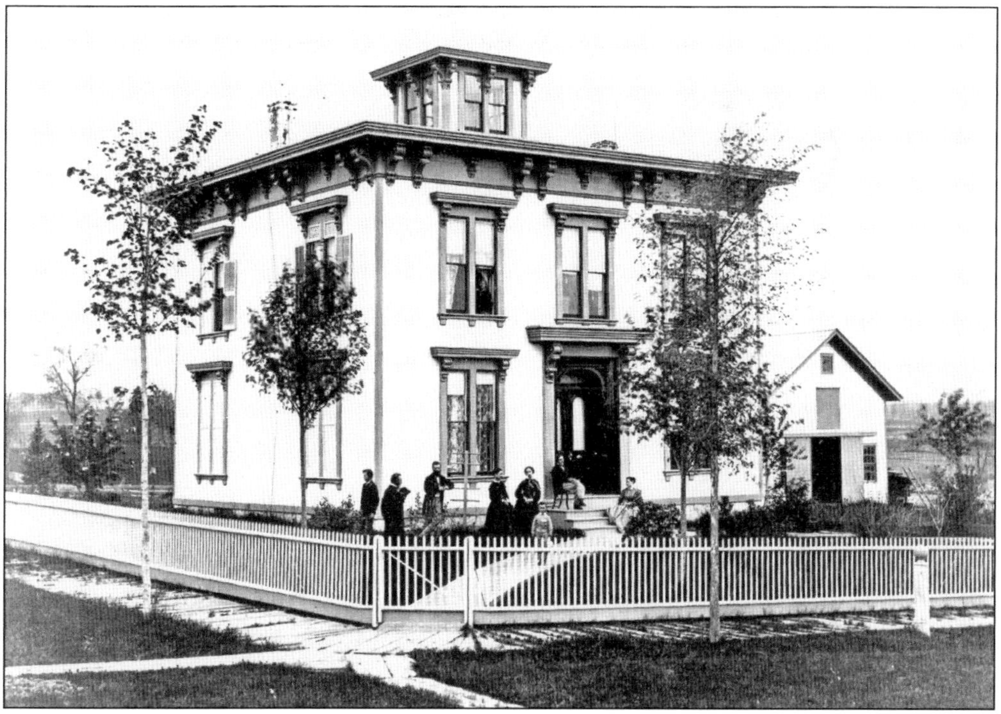

John M. Whipple built this Italianate Revival home at the corner of Summit and Erwin Streets in 1857. Like his uncle Squire Whipple, John Whipple was an engineer and bridge designer.

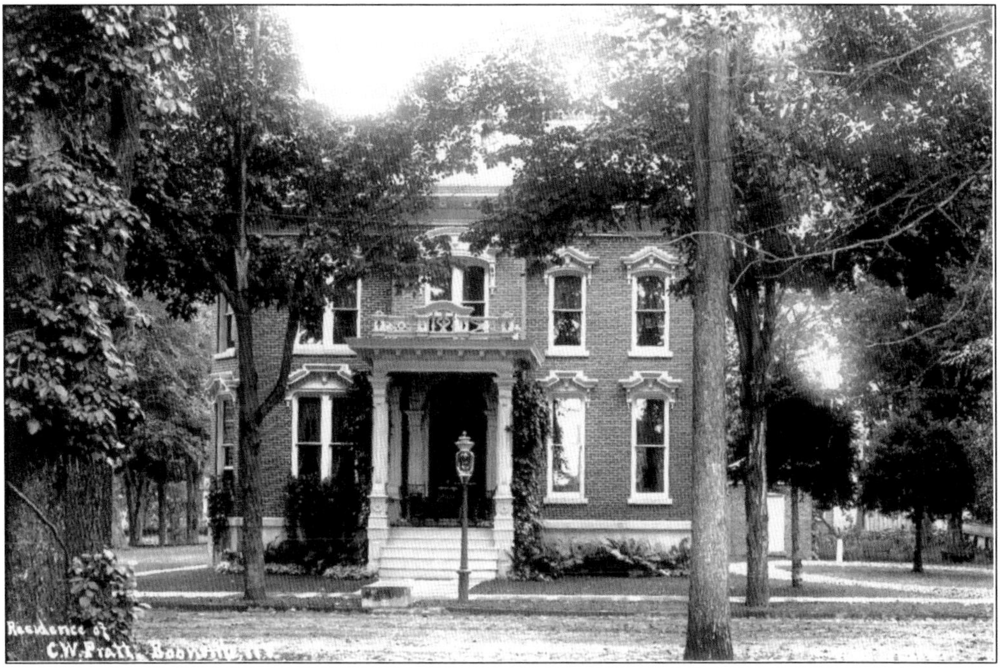


Charles W. Pratt, a principal in the Gould Paper Company, purchased this Schuyler Street house in the 1890s. The house, with its Second Empire styling, was built around 1875 by Clark Dodge. Canal tolls were deposited at the Bank of Boonville, which Dodge helped found.

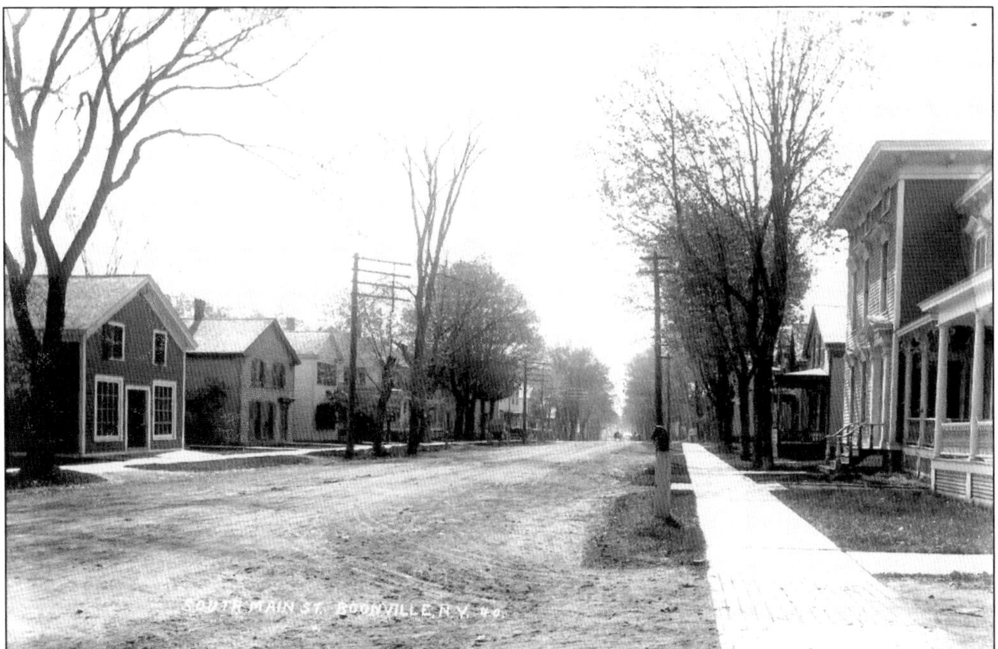


Houses on the streets bordering the canal tended to be modest. The view is of Main Street, looking southeast.

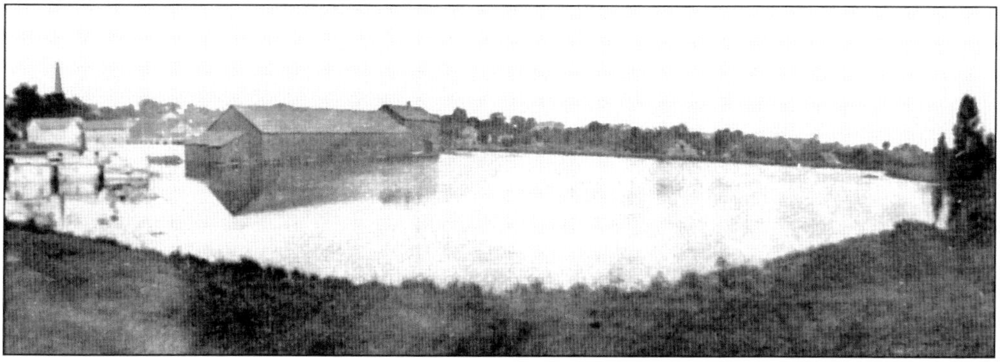

Canal boats are moored in the boat basin, now the municipal swimming pool, just south of the Erie Canal Feeder. The large building on the left is the state shop. The view is to the northwest.

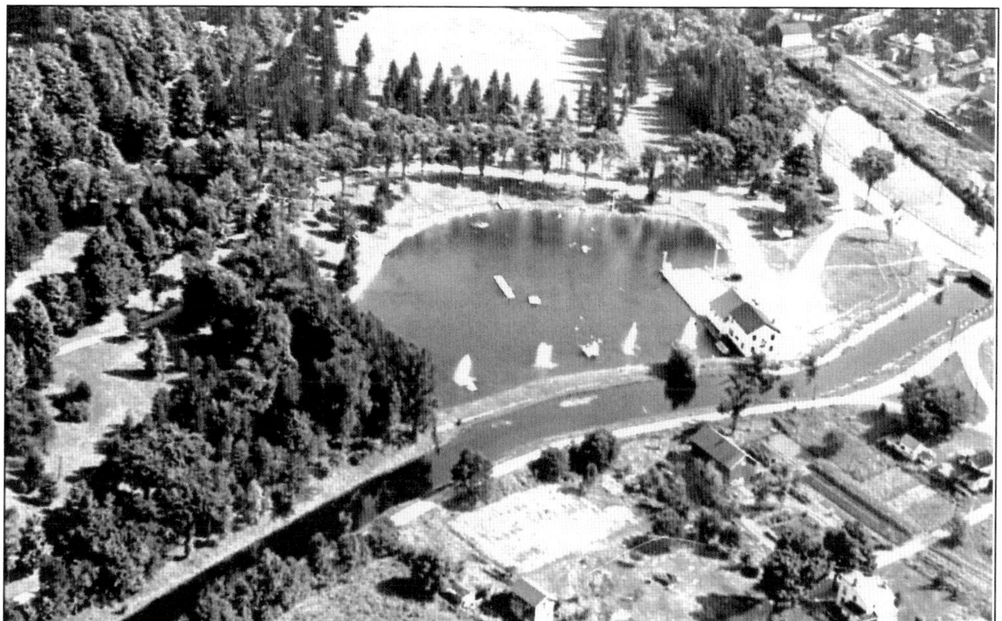

The boat basin, just east of the canal, became a municipal swimming pool in 1932 and is now part of Erwin Park. The swimming pool is supplied by Black River water pumped in from the Erie Canal Feeder, which runs along the berm behind the fountains. The building adjoining the swimming pool and the feeder was the state shop.

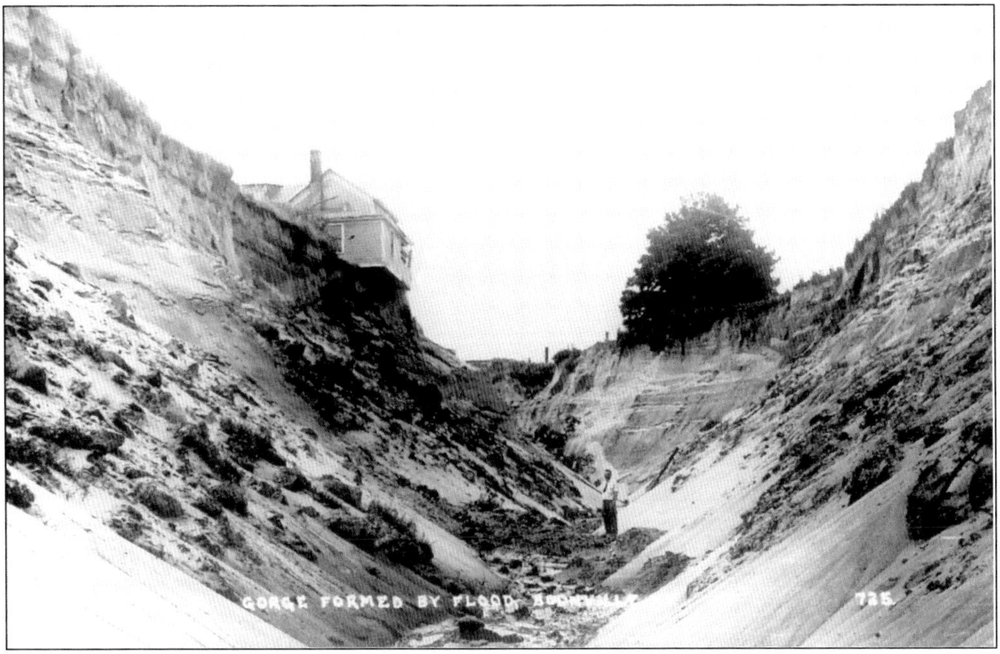


In building the swimming pool, workers inserted a pipe through the towpath wall of the feeder. As water leaked along the sides of the pipe, a trickle became a torrent. The towpath washed out, and the feeder emptied, gouging out a 75-foot-deep gully and burying East Schuyler Street with thick slurry. The teetering house belonged to Fred Duvelow.

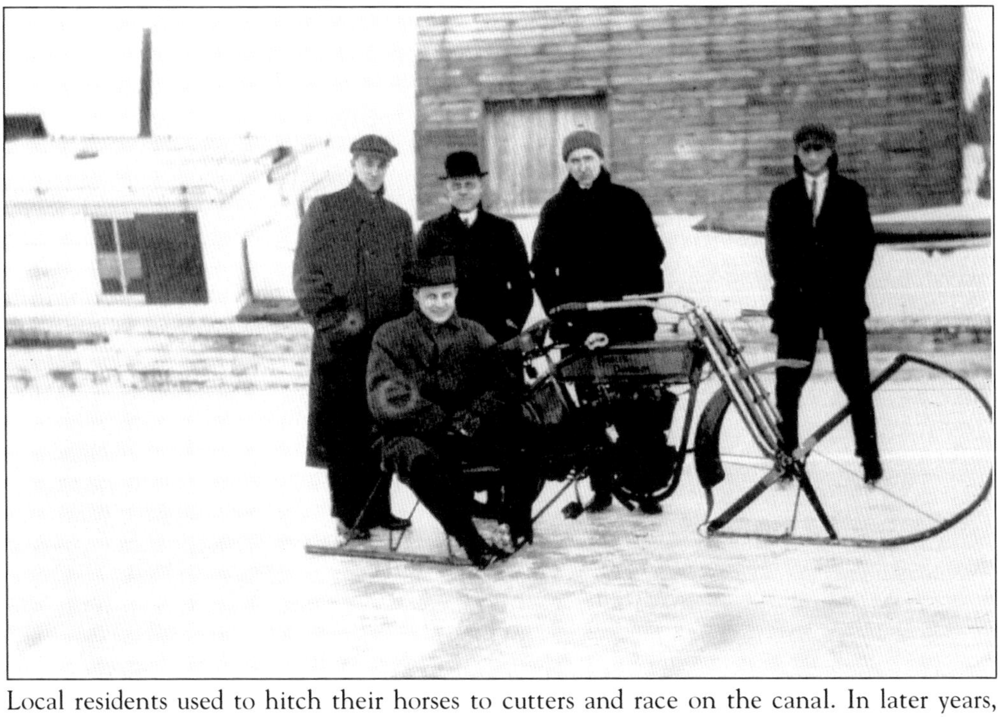

Local residents used to hitch their horses to cutters and race on the canal. In later years, Ed Buckley built a motorized "ice creeper," using a Harley-Davidson frame. (Courtesy of New York State Museum.)

# *Five*

# Northern Canal

The northern section of the overland canal opened as far as Port Leyden in 1851 and as far as Lyons Falls in 1855. The gradient of the northern section was much less than that of the southern section. Nonetheless, between Boonville and Lyons Falls, the northern canal descended 386 feet and required 39 locks. For the first four miles, the bedrock was limestone and, for the last six miles, it was metamorphic rock typical of the Adirondack shield. The most significant engineering challenge was at Lyons Falls. The canal had to be built around the high falls, and a series of seven locks was needed to lower boats into the Black River.

From Lyons Falls to the canal's terminus at Carthage, a distance of some 42 miles, the Black River itself was to be improved for slack-water navigation. Steamboats, which would ply this stretch of the river, required a channel that would accommodate canal boats with a four-foot draft. At first, engineers had planned to eliminate sandbars by constructing jetty dams (parallel rows of pilings that narrowed the channel and thereby generated a faster current). When legal difficulties halted work in 1852, the Canal Board decided to install a dam and lock at Otter Creek and one at Bush's Landing, east of Lowville. But when fraud was discovered in letting the contracts for this work, the Canal Board in 1854 returned to the idea of jetty dams. Soon, stream dredges and pile drivers were at work. In 1857, however, the Canal Board reversed itself, opting once again for locks and dams. The dam at Otter Creek was completed in 1860, and one at Bush's Landing was completed in 1866.

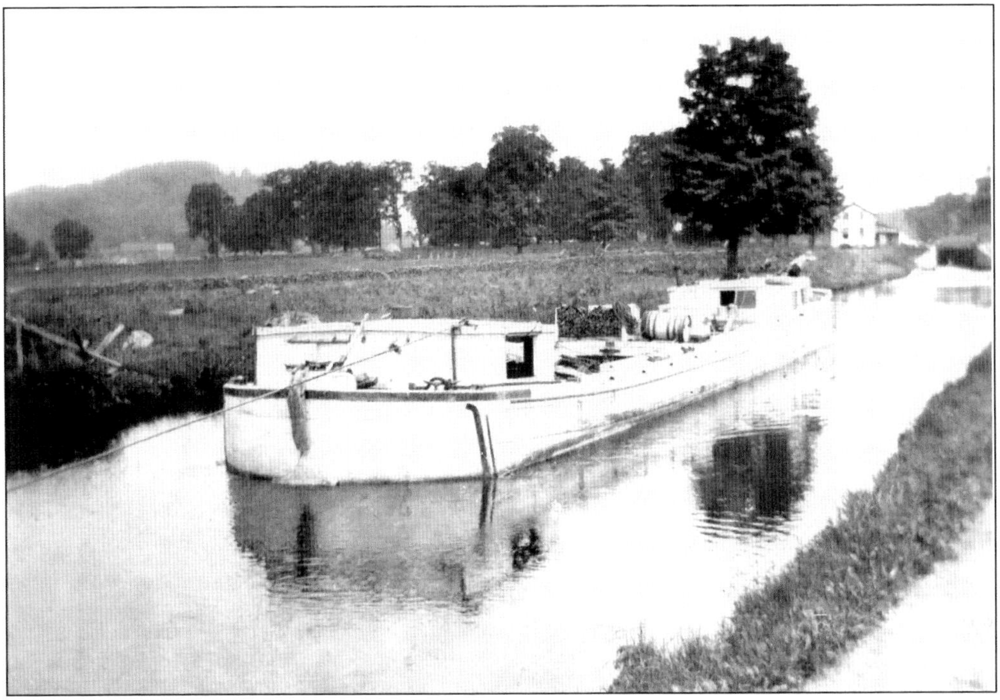

Beginning in 1862, all canal boats (with the exception of state scows) were required to have rounded prows to prevent them from catching on other boats, locks, or objects protruding from the banks. In the canal's heyday, dozens of boats might have been on the waterway at any one time. (Courtesy of Vincent T. Fitch.)

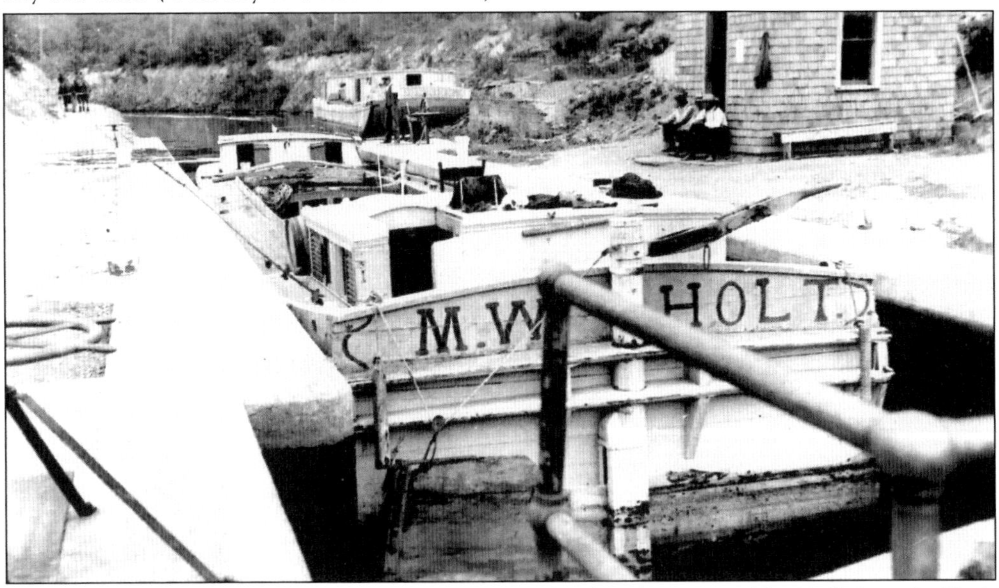


The M. W. *Holt* of Port Leyden was built at Castorland and captained by Dell Satterly. Each lock had two snubbing posts on the towpath side. To stop a moving boat, the driver used a hawser attached to the vessel. The driver would first wrap the line around one post and then around a cleat on the side of the boat, all the while drawing the line tight. The procedure was repeated at the second post.

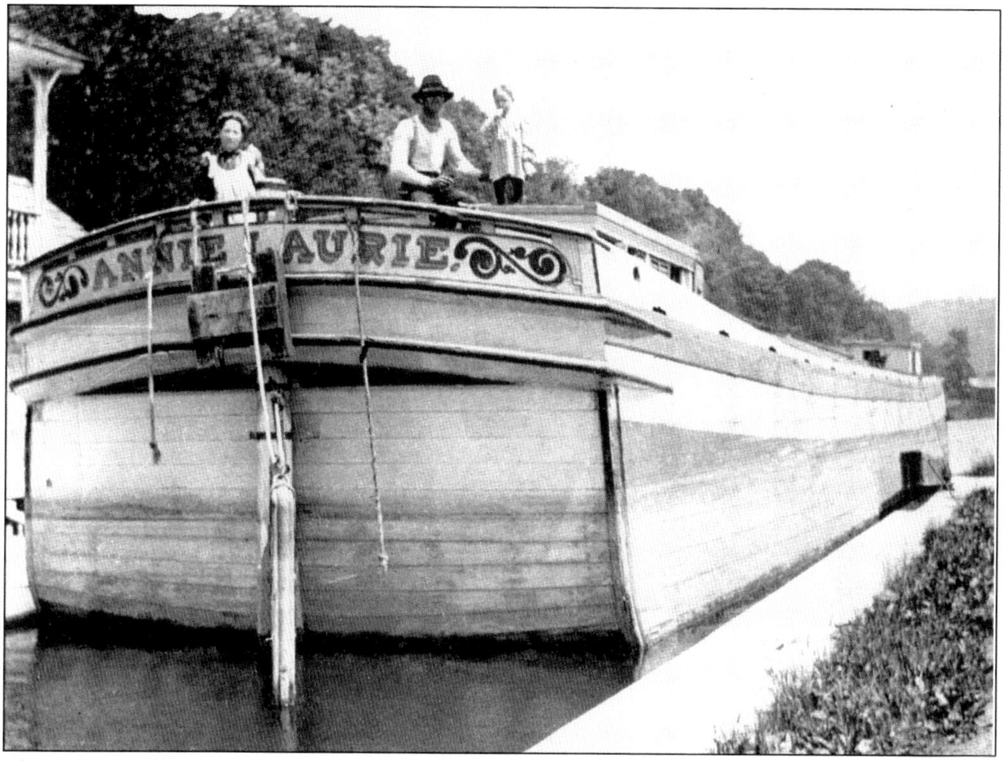


The captain of the *Annie Laurie* was Arthur "Archie" Lane of Port Leyden, shown here with his wife, Celia, and son, Walter. Small children were often tethered to keep them from falling off the boat and drowning. Despite the hardships of canal life, captaincy was often handed down from father to son.

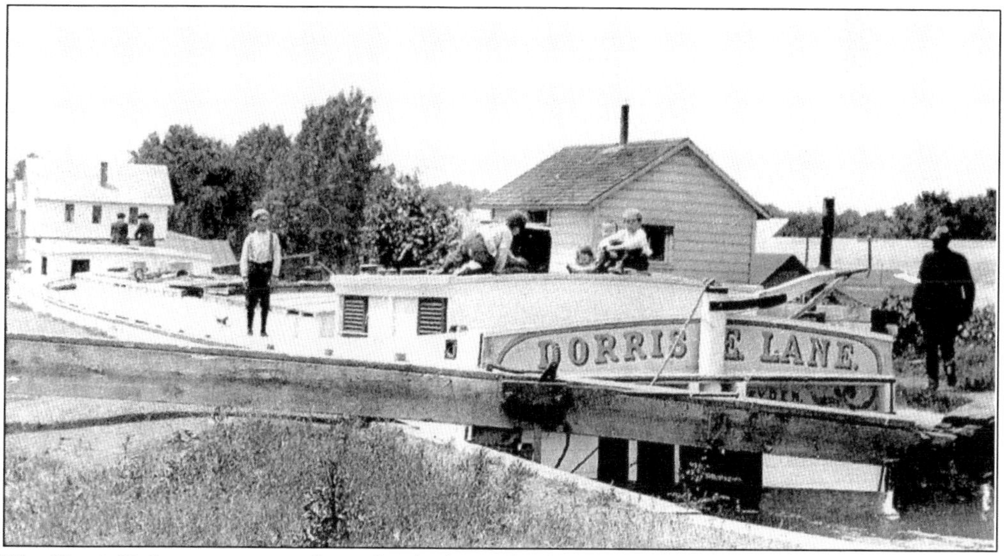


The *Dorris E. Lane* is passing through a lock north of Sugar River. Canal families lived in a stern cabin measuring about 9 feet by 12 feet. The interior was lit by kerosene lantern and was fitted out with a cookstove, kitchen table and stools, rocking chair, linoleum floor, and curtains. Sleeping quarters were behind and beneath the cabin. (Courtesy of New York State Museum.)

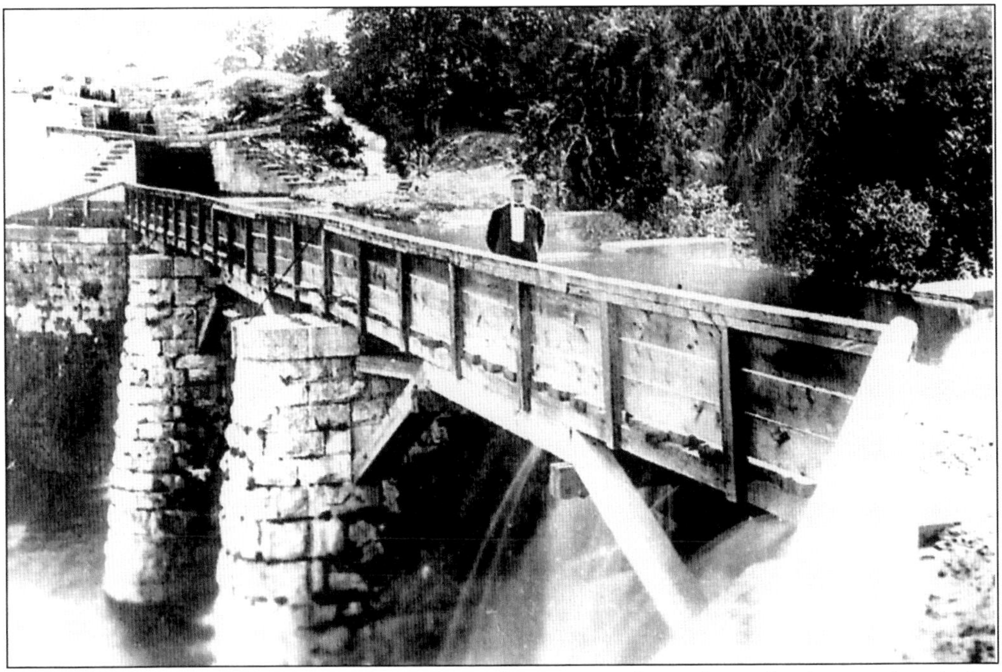

The canal was extended across Sugar River by means of an elevated, if somewhat leaky, aqueduct. At the left are adjoining Locks 81 to 84, excavated from bedrock. A short feeder at the head of Lock 84 replenished the canal with water from Sugar River.

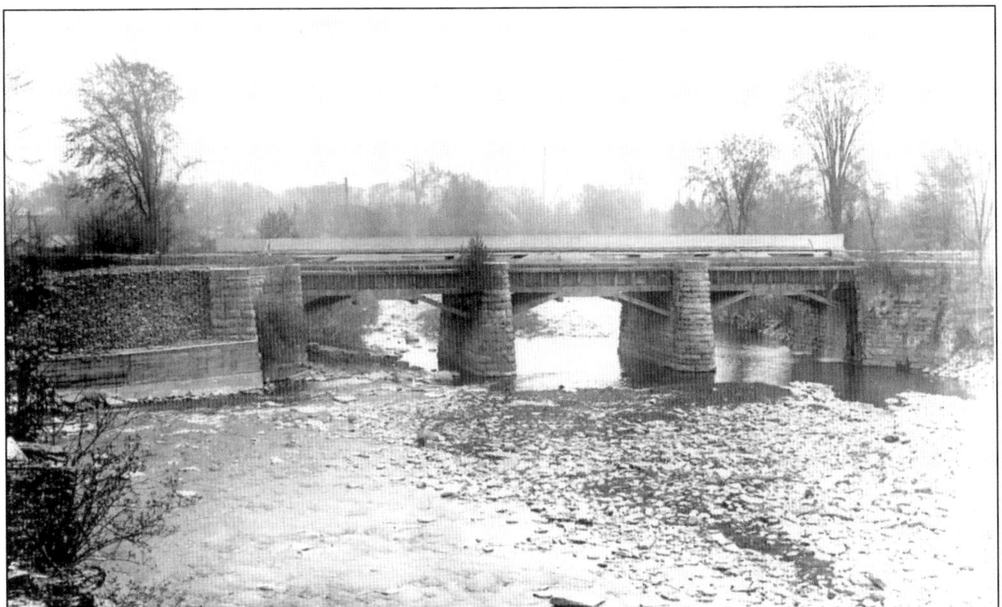

The Sugar River aqueduct had three 40-foot wooden arches and stone piers 5 feet 11 inches wide. The view is toward the east. In the distance is the Sugar River quarry.

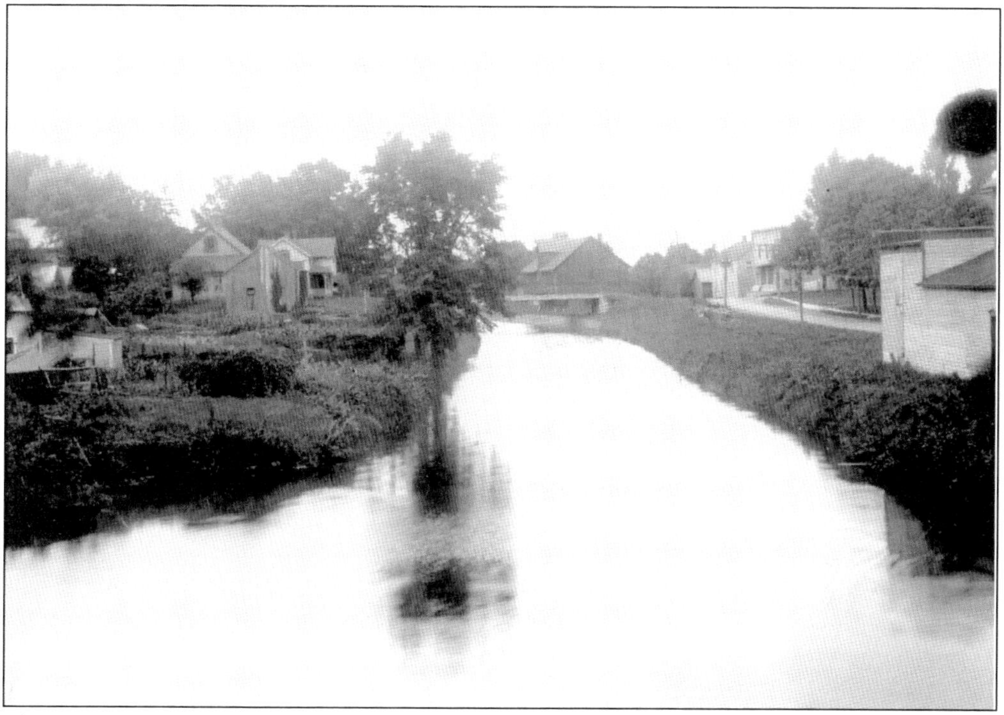

The canal runs past a thriving household garden not far from Lock 96 and the Main Street bridge in Port Leyden. The canal had been abandoned when this photograph was taken. (Courtesy of Port Leyden Community Library.)

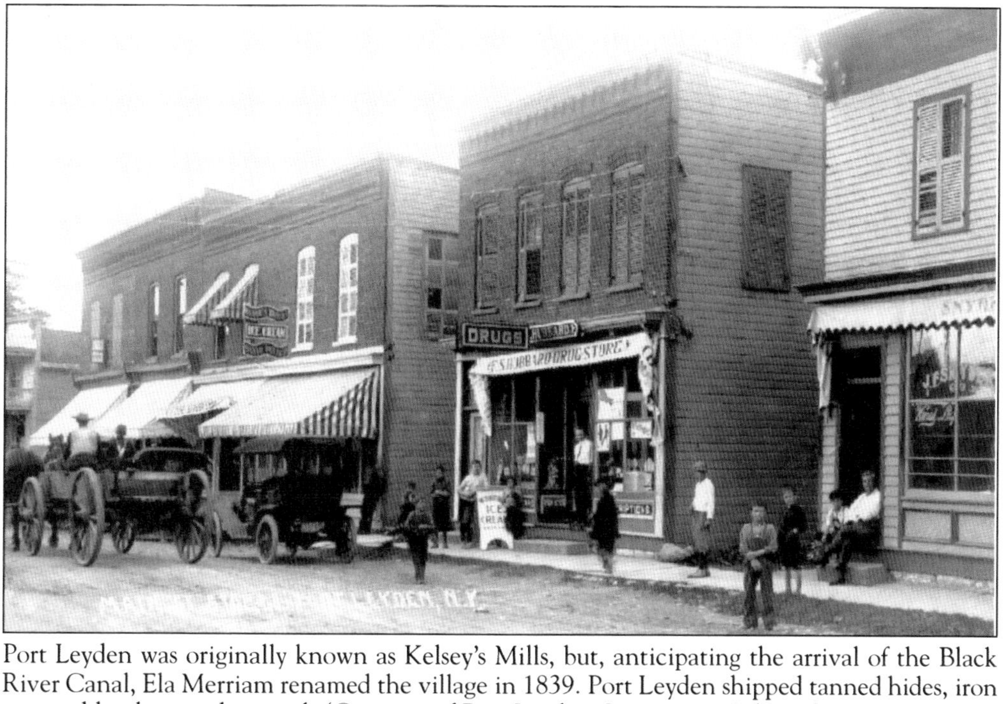


Port Leyden was originally known as Kelsey's Mills, but, anticipating the arrival of the Black River Canal, Ela Merriam renamed the village in 1839. Port Leyden shipped tanned hides, iron ore, and lumber on the canal. (Courtesy of Port Leyden Community Library.)

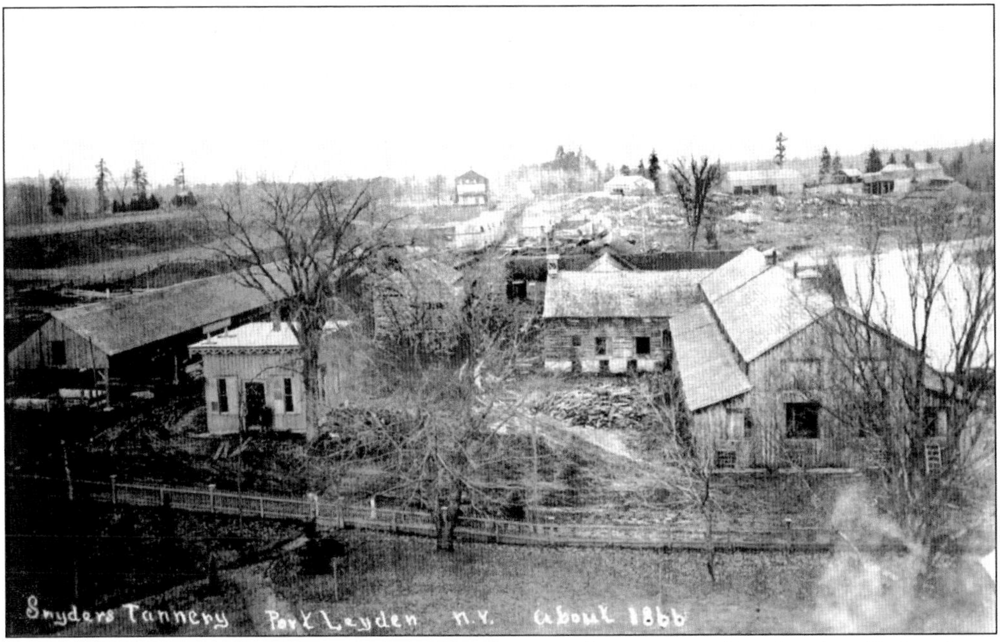


In 1855, H. D. H. and Augustus Snyder bought a small tannery at Port Leyden. Before long, the tannery employed 300 persons and was turning out 40,000 hides of sole leather a year. The tannery burned down in 1875. (Courtesy of Port Leyden Community Library.)

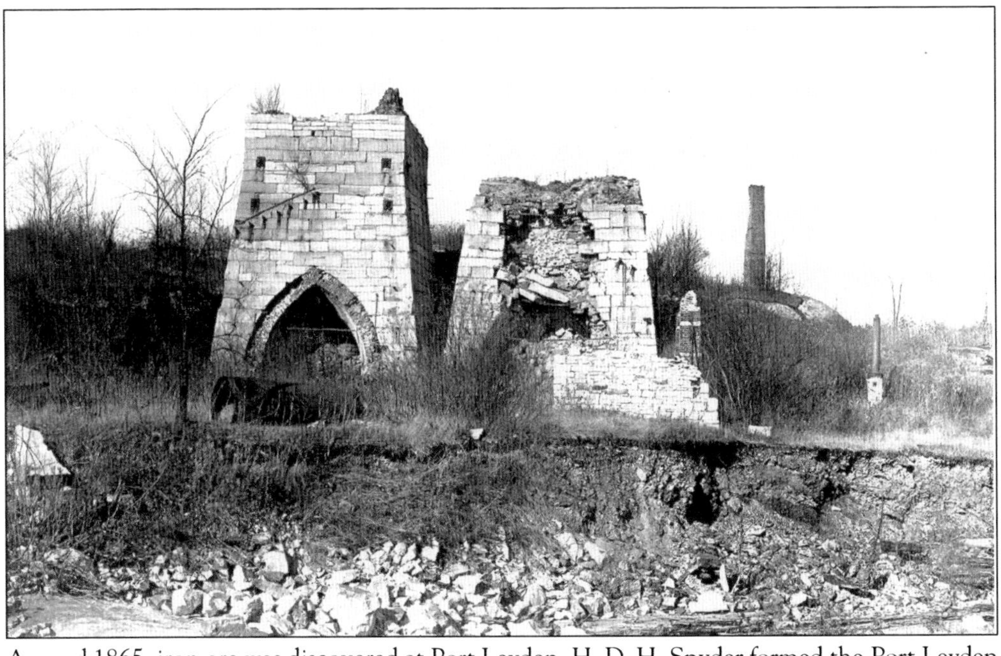

Around 1865, iron ore was discovered at Port Leyden. H. D. H. Snyder formed the Port Leyden Iron Company to extract and refine the ore. The enterprise later passed into other hands, but a lack of high-grade ore prevented the business from flourishing. The furnaces were known as "Grace" and "Fannie." (Courtesy of Port Leyden Community Library.)

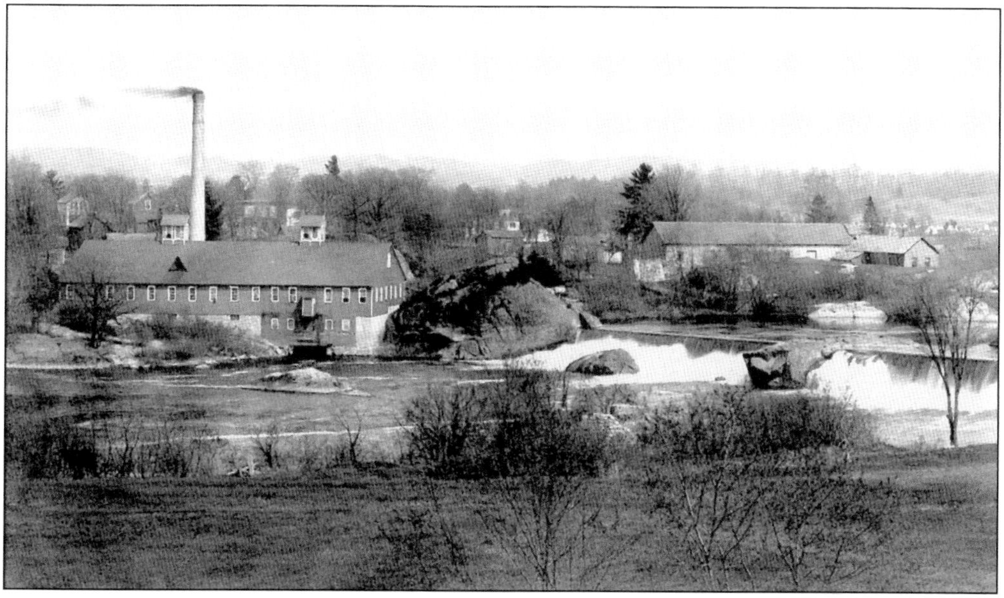

Besides his pulp mill at Port Leyden, G. Henry P. Gould had mills at Gouldtown, Fowlersville, Shuetown, Kosterville, McKeever, and Lyons Falls. Founded in 1892, the Gould Paper Company acquired vast tracts of forest in the Moose River region and on Tug Hill. Although local mills consumed huge quantities of pulpwood, a great deal of this softwood was also exported on the canal. In terms of annual tonnage, pulpwood was typically second only to lumber. (Courtesy of Vivian Seeloff.)

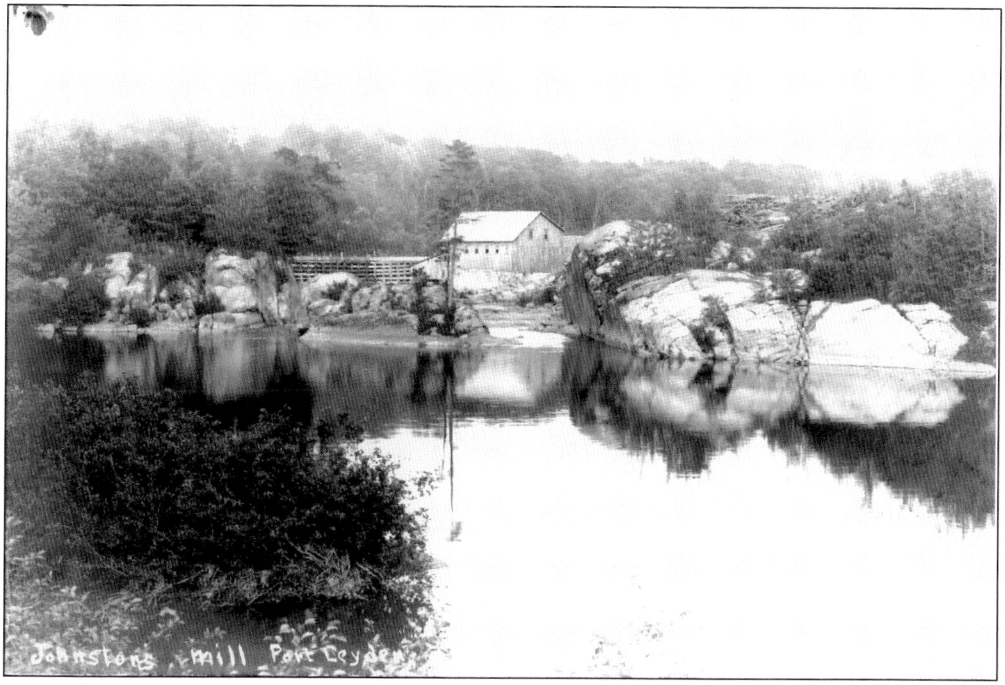


Dr. William H. Johnston's pulp mill at Port Leyden was built in 1896. Taken over by his son John E. Johnston, the mill remained a going concern until 1933. (Courtesy of Port Leyden Community Library.)

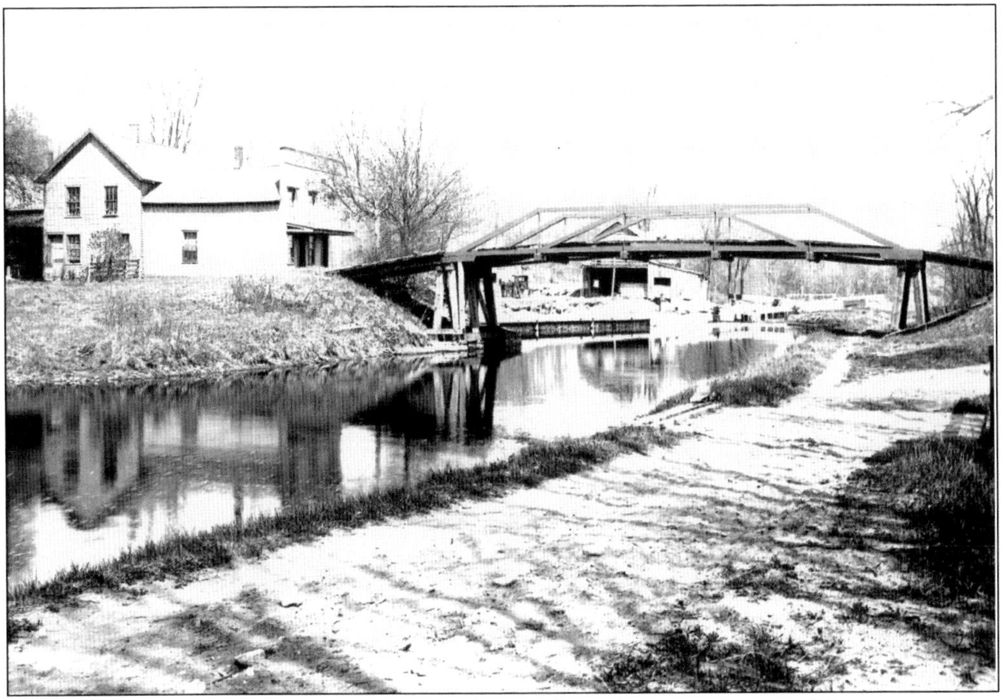

The northern canal had several "change bridges," where the towpath crossed from one side to the other.

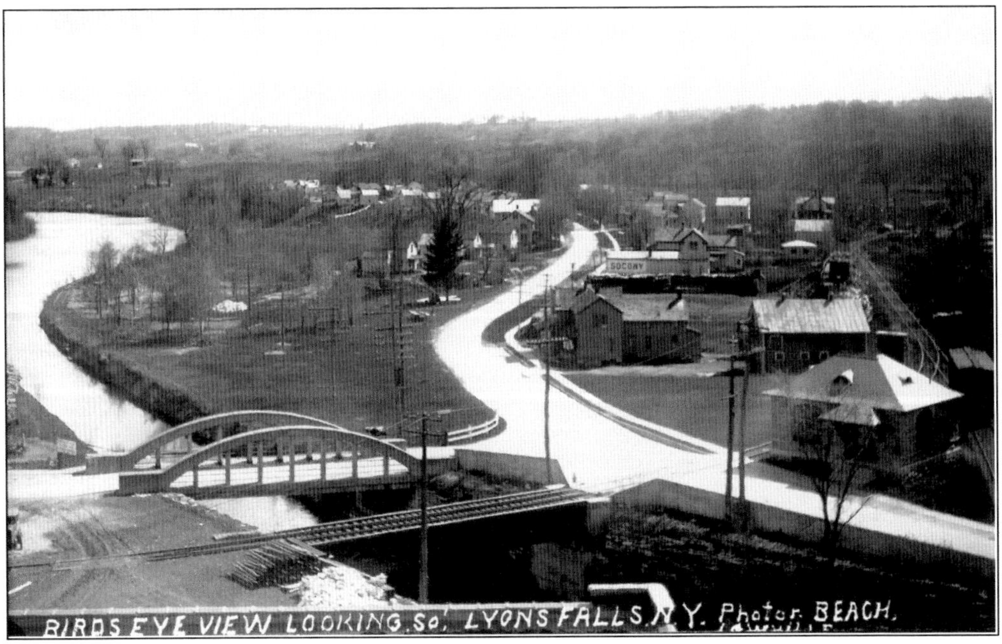


The canal entered Black River just south of Lyons Falls and then detoured around the high falls. The road bridge, pictured here, was removed around 1946. (Todd Weseloh Collection.)

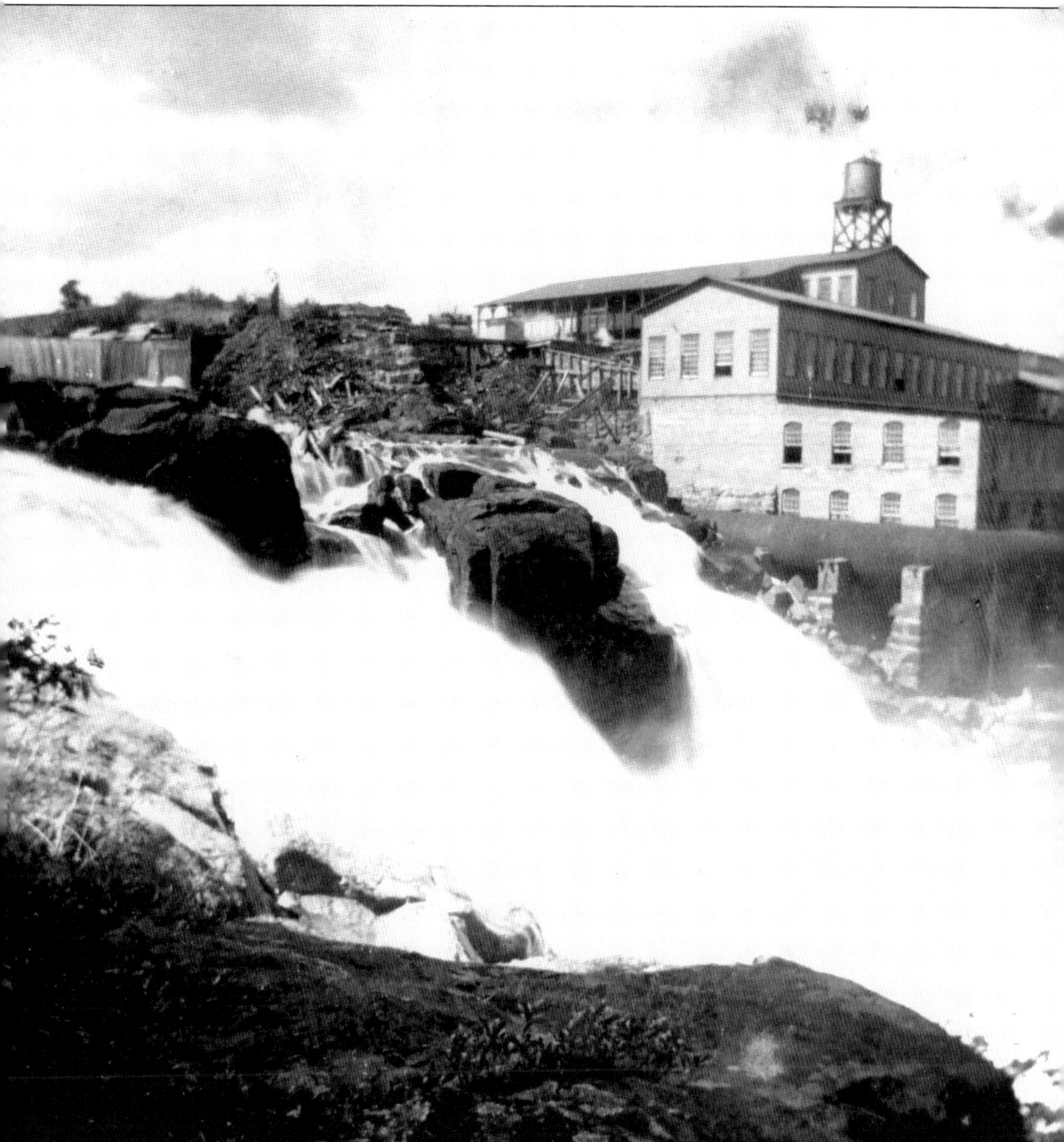

Located between Lock 102 and Lock 103, the 70-foot high cataract at Lyons Falls represented the greatest obstacle in building the northern canal. The towpath from Lock 102 runs along the west bank to the side-cut around the falls and descends to the river through seven locks.

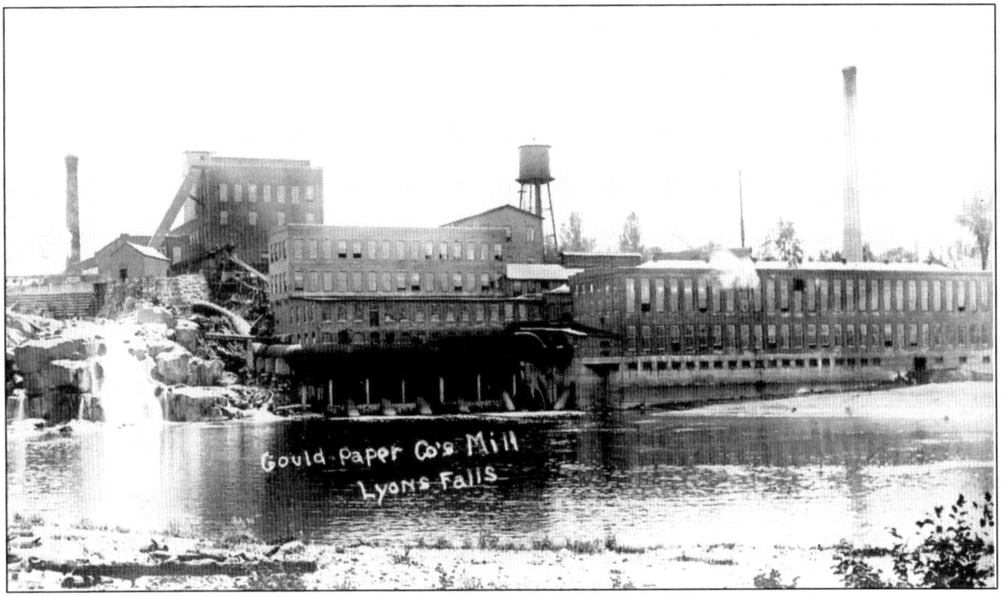


G. H. P. Gould built a huge paper mill at Lyons Falls in 1895. It was the first in the United States to use electricity. Brick for the mill was imported on the canal.

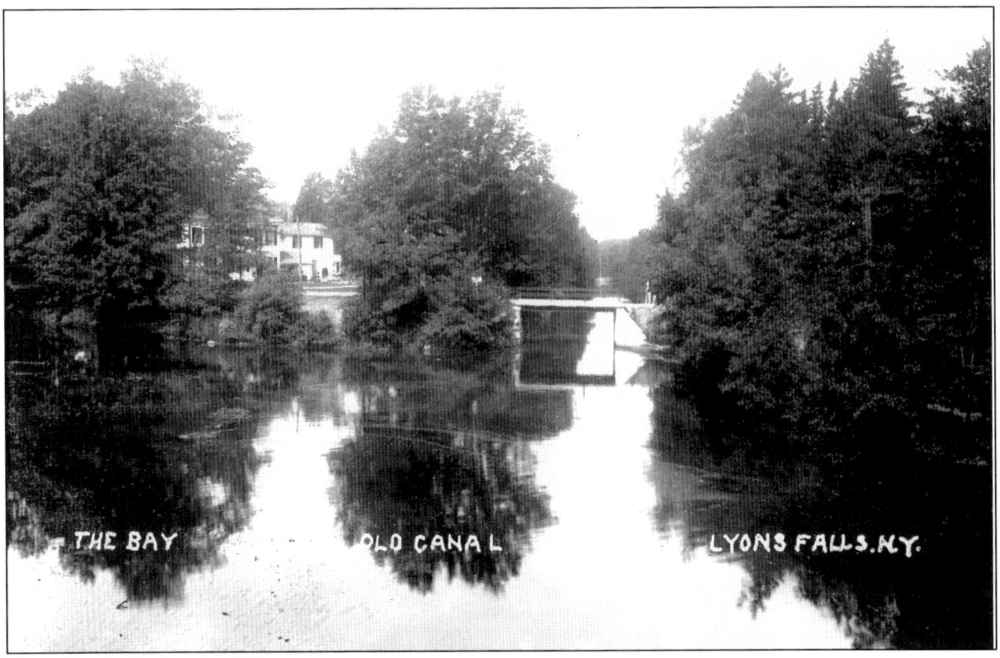


The view is looking north from the Lyons Falls boat basin toward the last locks on the overland canal. In the heyday of the canal, dozens of boats might be moored here at any one time. The canal was no longer in operation when this photograph was taken. (Courtesy of George Cataldo.)

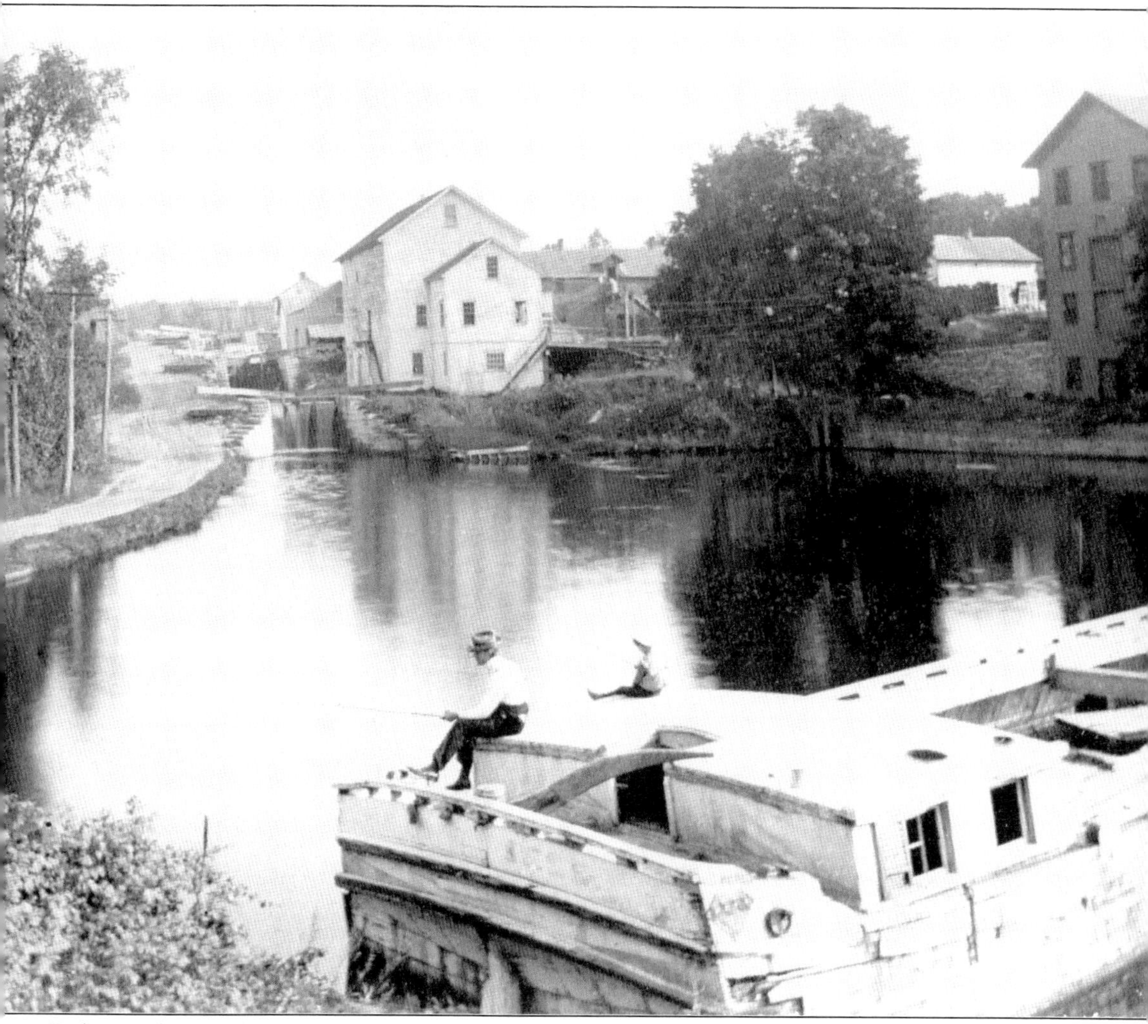

Father and son fish from a canal boat in the Lyons Falls basin, hoping perhaps to catch one of the canal's legendary seven-pound eels. Boats unloaded dry goods at Watson Shaw's general store, the four-story building on the right. Shaw built this store in 1895 and operated it until 1914. The building in the center is Homer Markham's feed mill, built in 1880. The mill was powered by canal water running through a sluice.

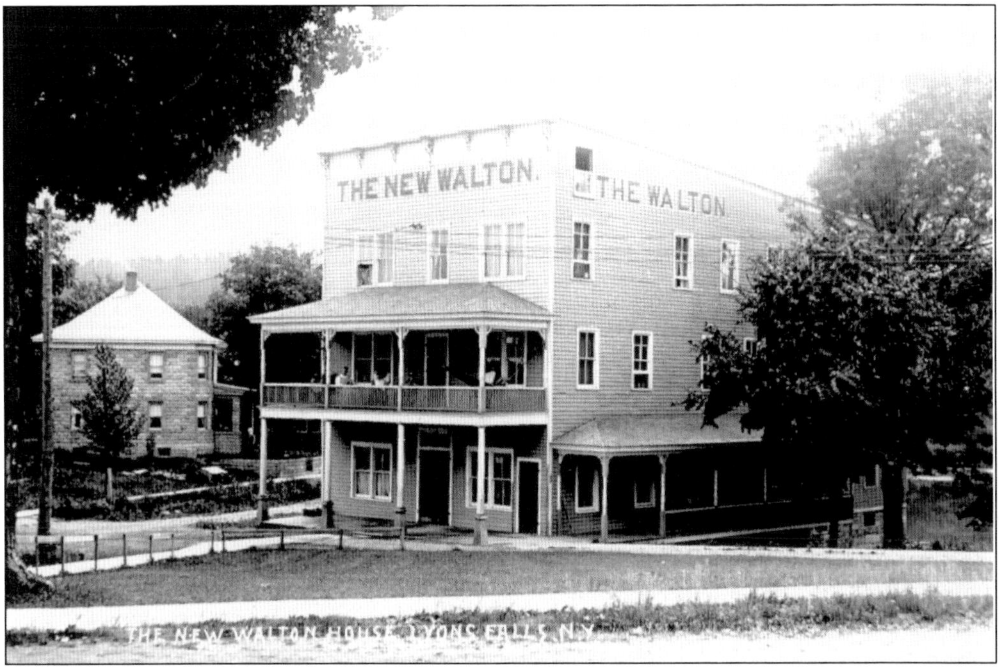


In 1907, Watson Shaw bought Archibald McVickar's general store and converted it into the New Walton Hotel, which later became the Lyons Inn. Built in 1867, this structure was located on the boat basin. Opposite the New Walton was the original Walton, a summer boardinghouse built by Gordias Gould in 1859. The Walton burned in 1901. (Courtesy of George Davis.)

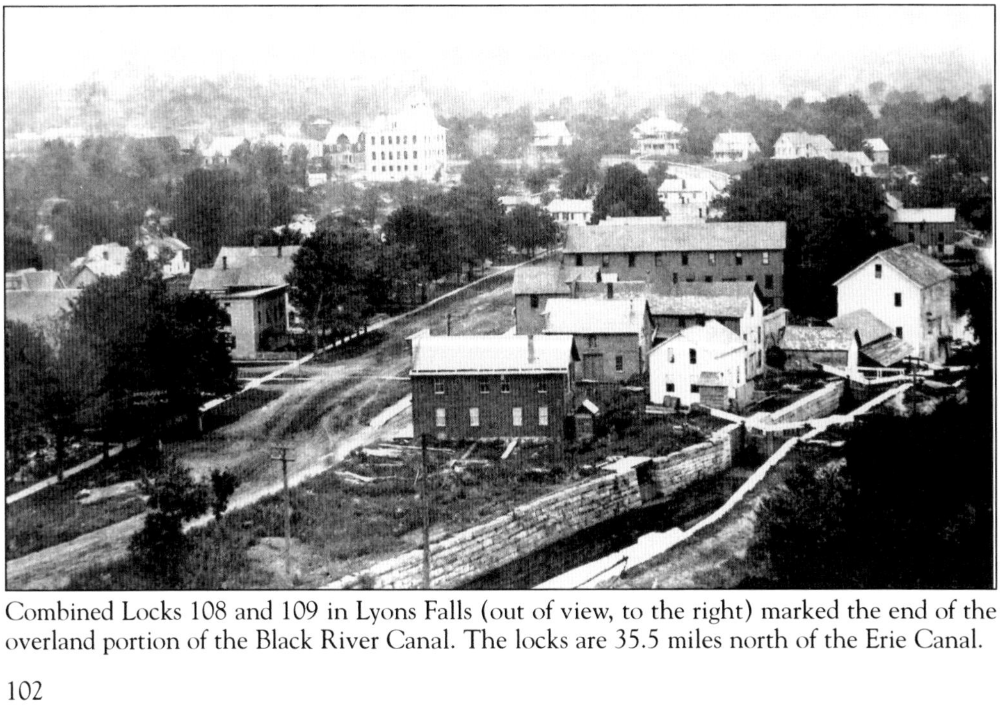

Combined Locks 108 and 109 in Lyons Falls (out of view, to the right) marked the end of the overland portion of the Black River Canal. The locks are 35.5 miles north of the Erie Canal.

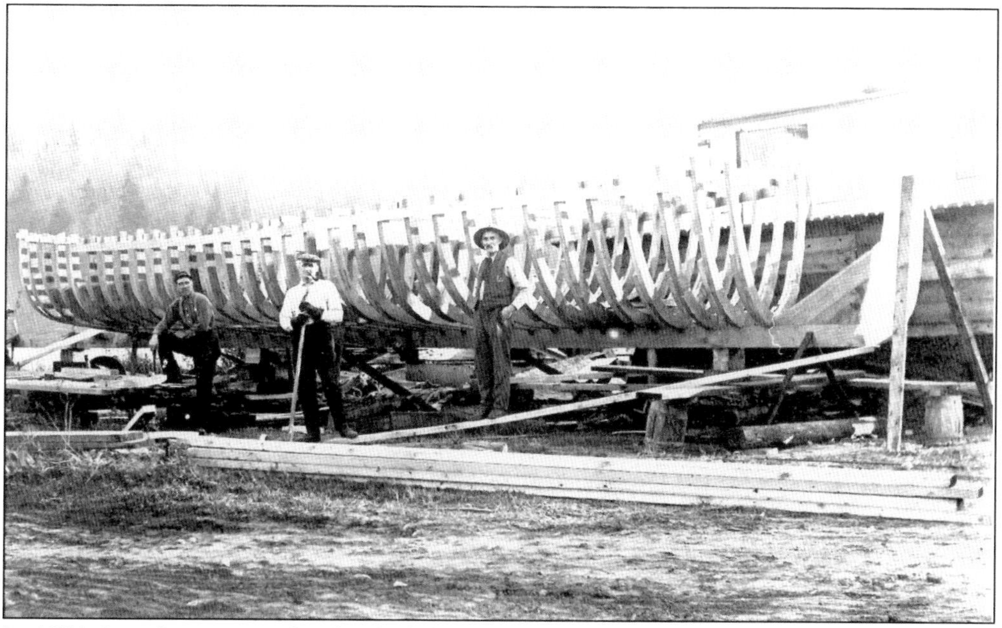

In 1876, Jesse Irons opened a dry dock, boatyard, and blacksmith shop on Abel Basin below Locks 108 and 109. (Courtesy of Mary Teal.)

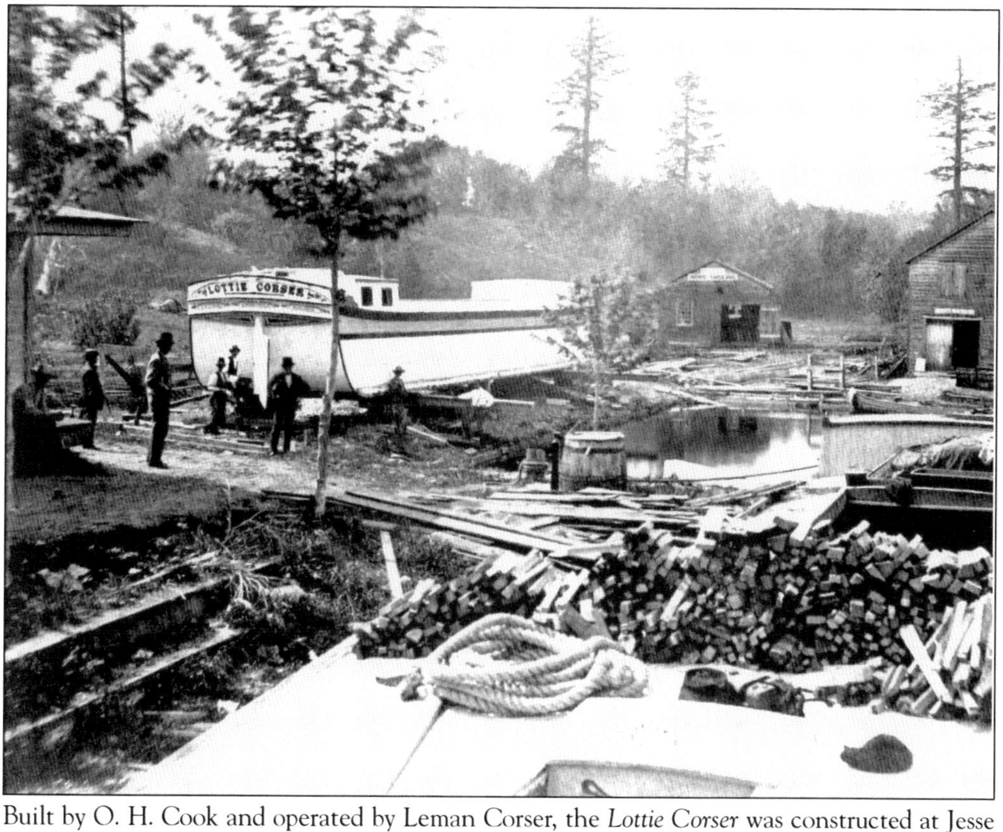


Built by O. H. Cook and operated by Leman Corser, the *Lottie Corser* was constructed at Jesse Irons's boatyard. (Courtesy of Mary Teal.)

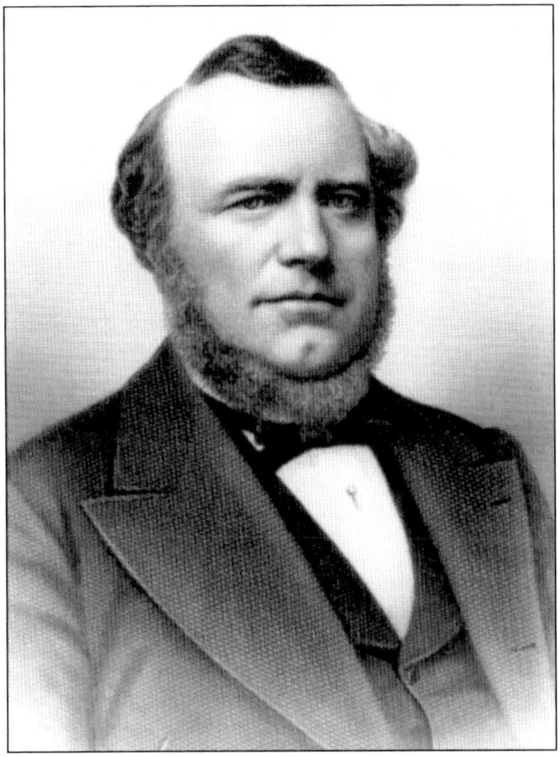

Lewis County landholder Lyman Rasselas Lyon formed the Black River Steamboat Company in 1856. That year, he launched an $8,000 sternwheeler, the *L. R. Lyon*, to tow canal boats between Lyons Falls and Carthage. The 132-foot steamer, captained by Walter Whittlesey, was built locally by George Baker and had two engines with an estimated 100 horsepower.

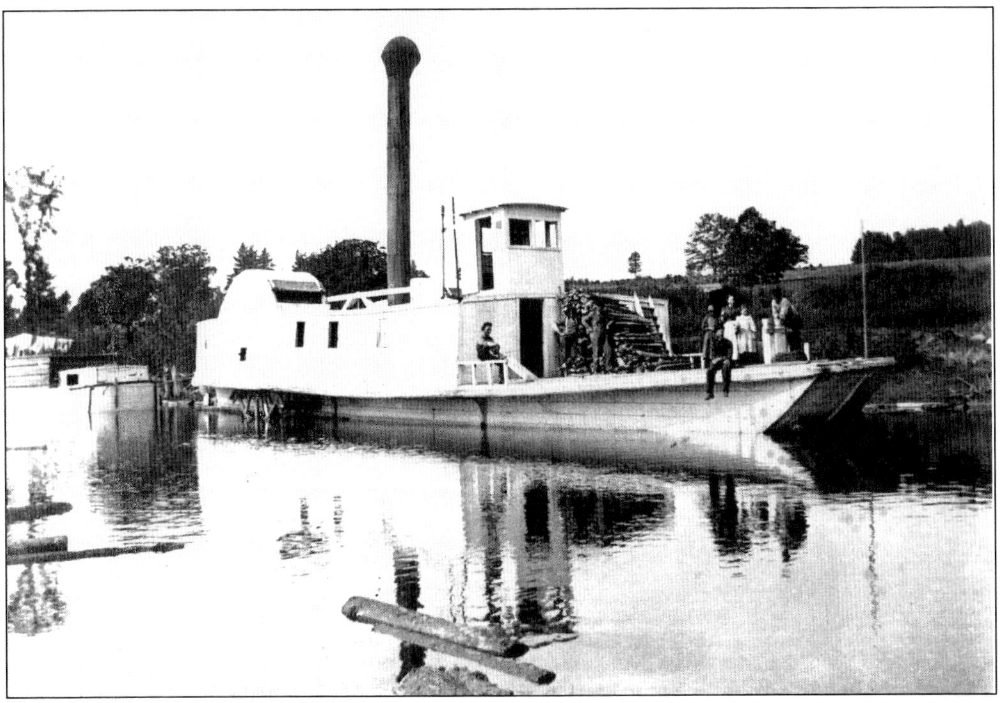

The *L. R. Lyon* was destroyed by fire in 1866. Its successor, the *L. R. Lyon II*, a side-wheeler, is pictured here towing Black River Canal boats. Notice the laundry drying in the breeze. The *Lyon II* burned at Lyons Falls in 1878.

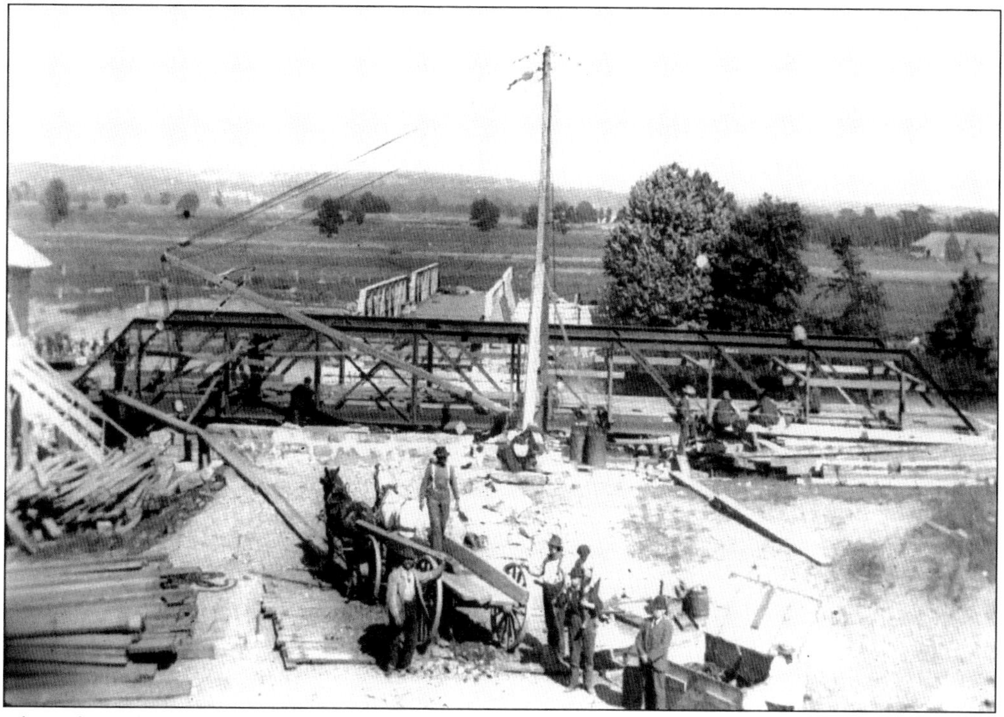

The Glenfield bridge is undergoing repairs in 1897. On the left is the Markham warehouse. Lyman Rasselas Lyon had the original covered bridge at Glenfield dismantled in 1856 so that his steamer could pass under it. (Courtesy of New York State Museum.)

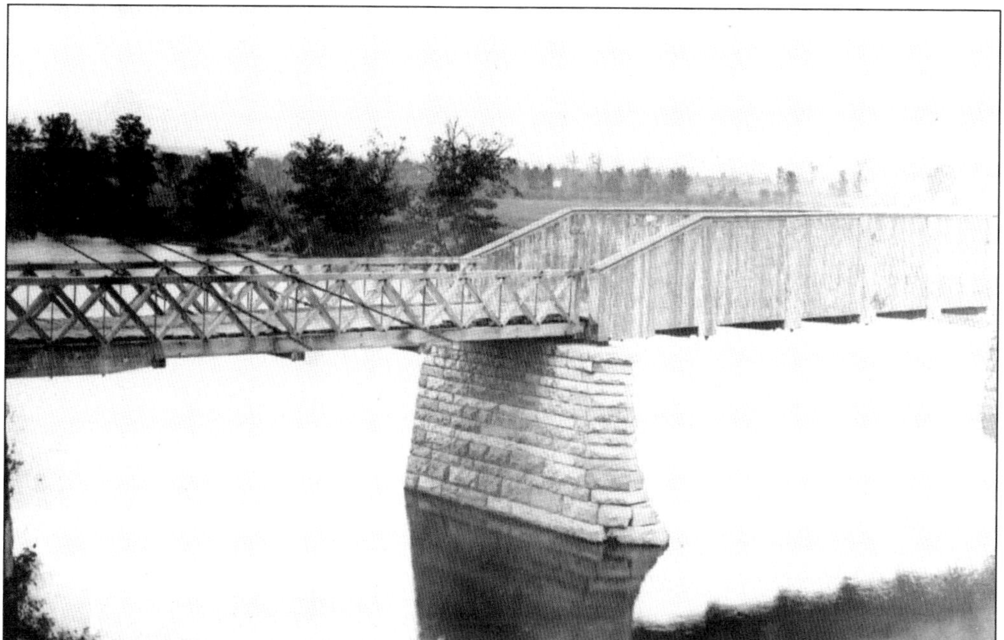

The Glenfield bridge was 231 feet long, with the western span functioning as a swing bridge. There were five other bridges on the canalized river: Whittlesey's (Lyons Falls), Beach's (near Lowville), Illingsworth (near New Bremen), Castorland, and Carthage. (Courtesy of Paul Joslyn.)

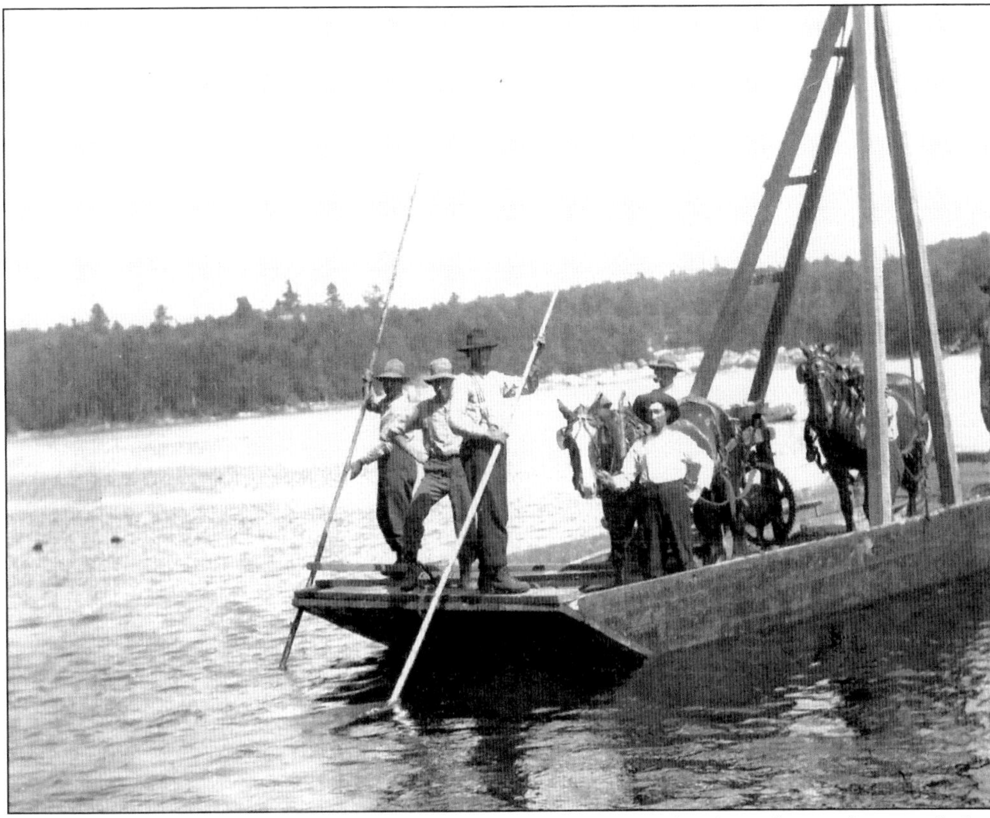

With only six bridges over a 42-mile stretch of canal in 1865, local residents often needed to use ferries to cross Black River. (Courtesy of Bob Hindman.)

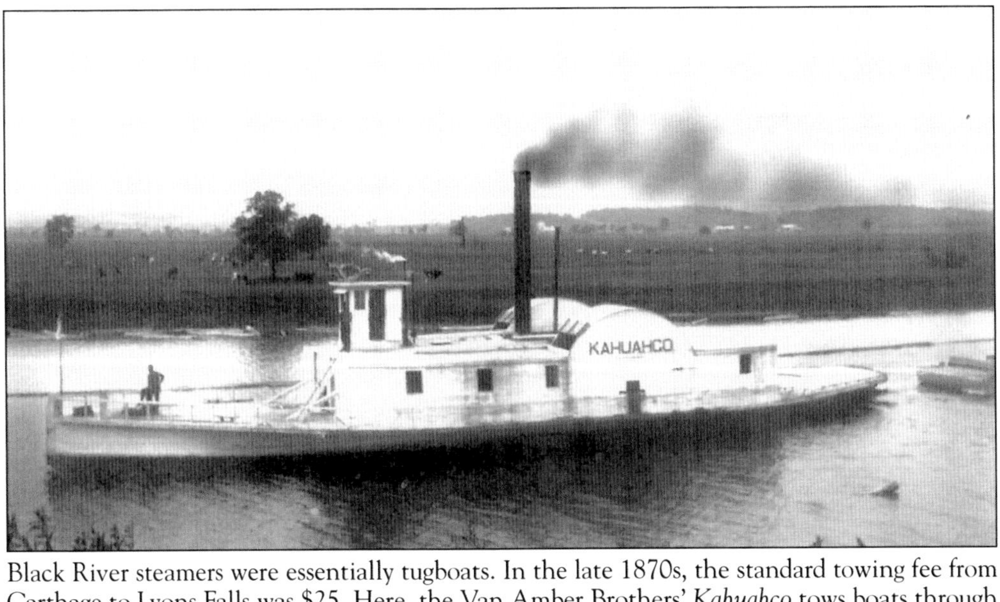


Black River steamers were essentially tugboats. In the late 1870s, the standard towing fee from Carthage to Lyons Falls was $25. Here, the Van Amber Brothers' *Kahuahco* tows boats through prosperous dairy country. Kahuahco was mistakenly thought to be the Native American word for "Black River." (Courtesy of Lewis County Historical Society.)

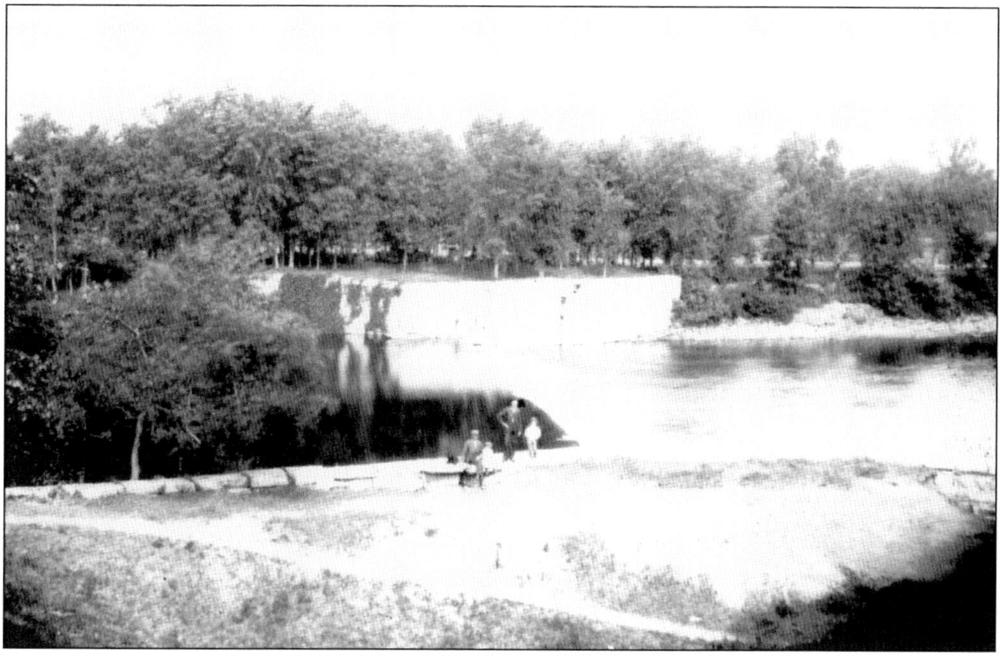

A low dam opposite the mouth of Otter Creek made slack-water navigation possible between Lyons Falls and Glenfield. Worringer's Grove is on the far side of the dam, and behind it is a lock that allows boats to bypass the dam. (Courtesy of Paul Joslyn.)

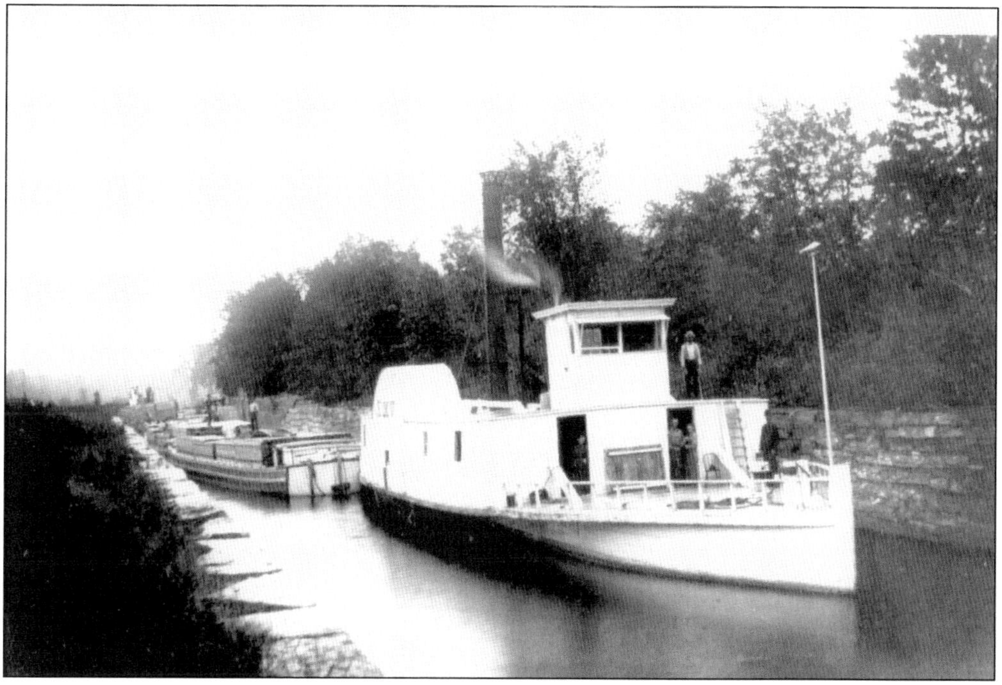

Named for Henry Van Amber's daughter, the *Edith M. Van Amber* is pulling lumber boats through the Otter Creek lock. Although steamboats had no standard size, they fit easily in the wooden locks at Otter Creek and Bush's Landing that measured 160 feet by 34 feet. (Courtesy of Paul Joslyn.)

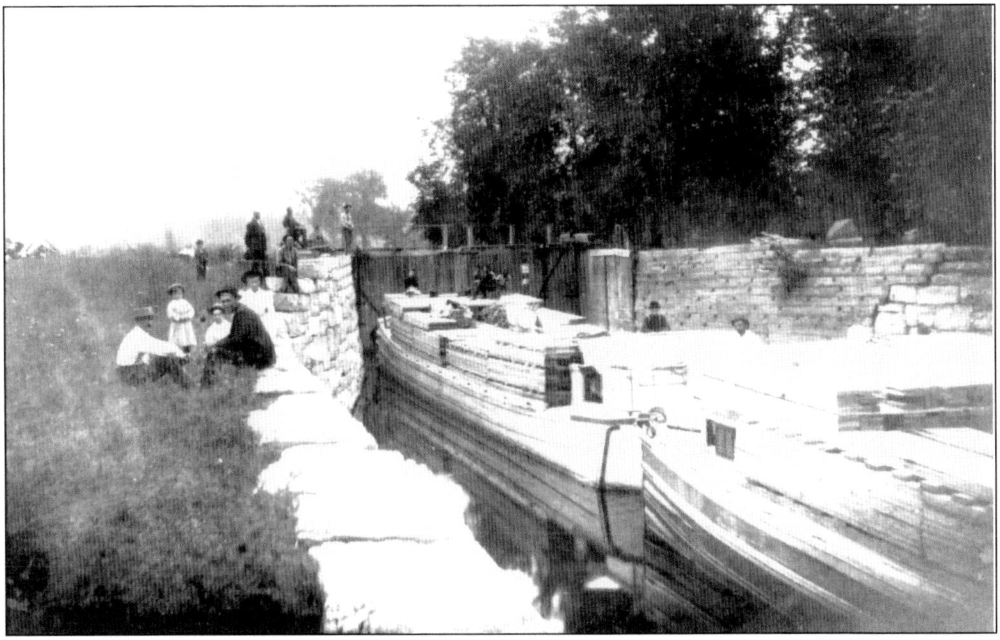

Van Amber canal boats are passing through the Otter Creek lock near Glenfield. Each boat typically carried 75,000 board feet of lumber. In the 1880s, a round-trip from Carthage to New York usually took three weeks and yielded a $350 profit for the boat owner. (Courtesy of Paul Joslyn.)

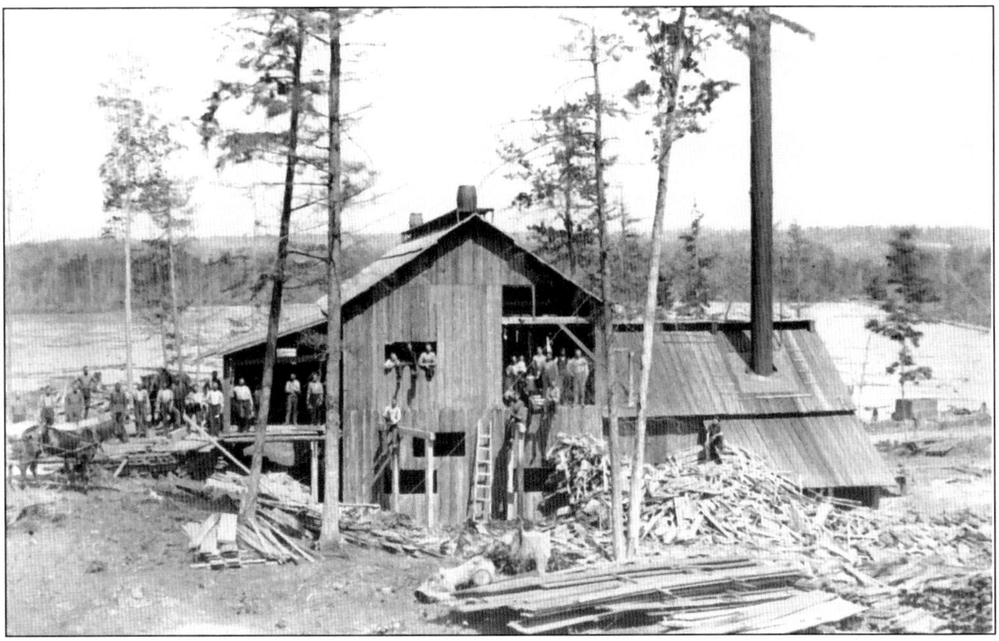

In 1885, Henry Abbey built a steam-powered sawmill near the mouth of the Independence River. The mill was capable of turning out seven million feet of lumber a year. Behind the mill, a 100-acre pond connects with the Independence. A three-quarter-mile narrow-gauge tram transported lumber to a loading dock on the canalized Black River. The photograph is from 1887, four years before the mill burned. (Courtesy of George Cataldo.)

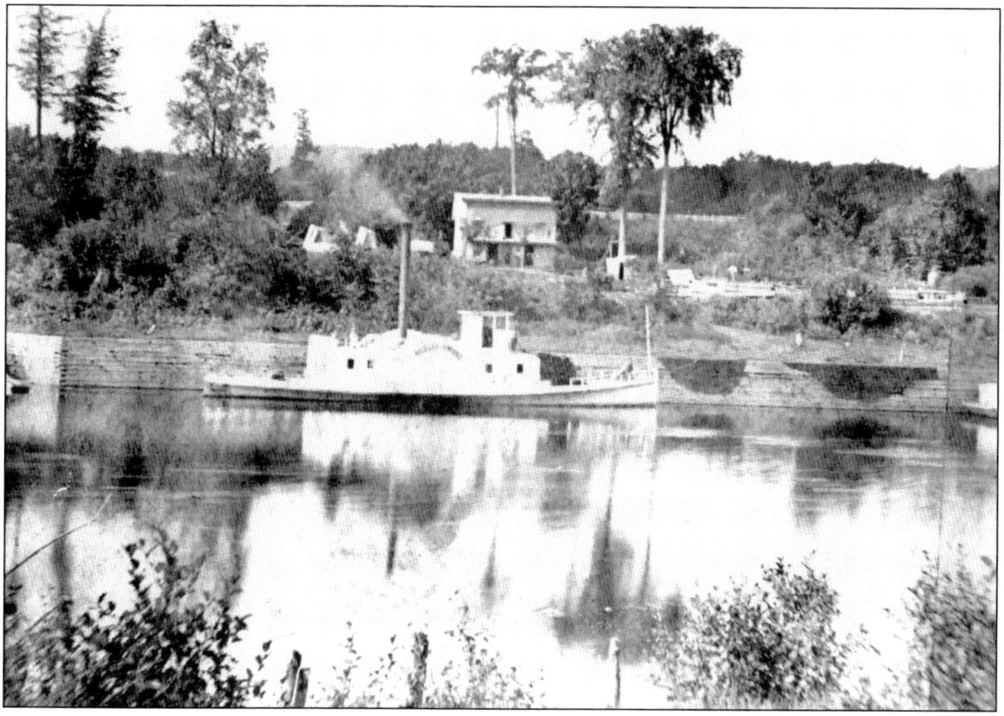

In the 1870s, Capt. George Sweet operated the *Nellie Sweet*, offering round-trip picnic parties between Carthage and Lyons Falls for a fee of $1.50 per person. The Carthage Cornet Band provided the music. The scene here is Bush's Landing. (Courtesy of Mary Teal.)

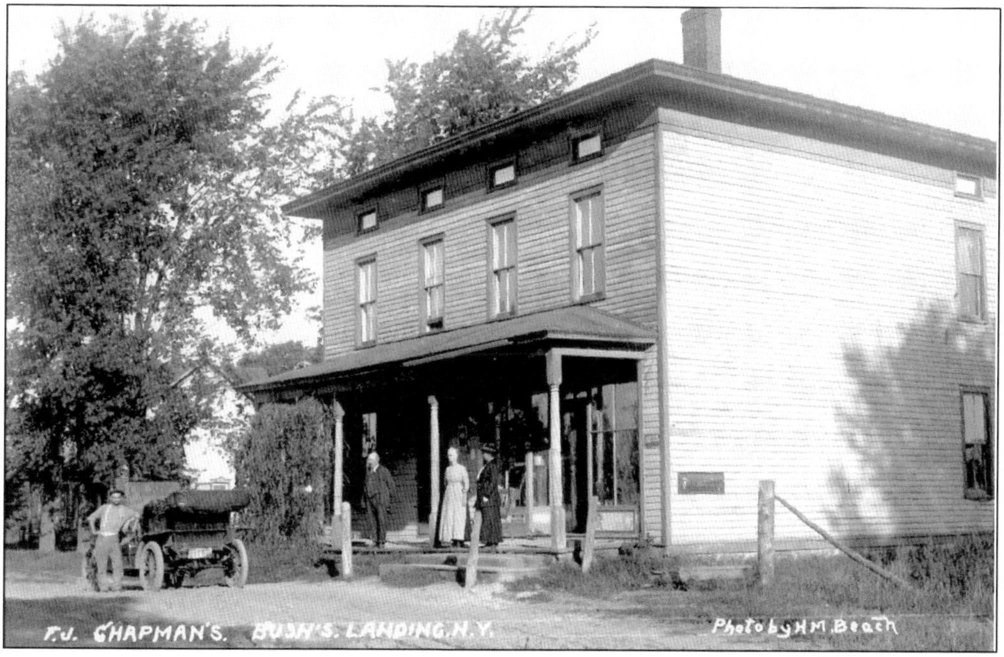


F. J. Chapman ran a dry goods store at Long Reach, or Bush's Landing, two miles north of the Independence River. Nearby, the Beach brothers, Andrew and Ralph, ran a sawmill and built canal boats. (Courtesy of George Davis.)

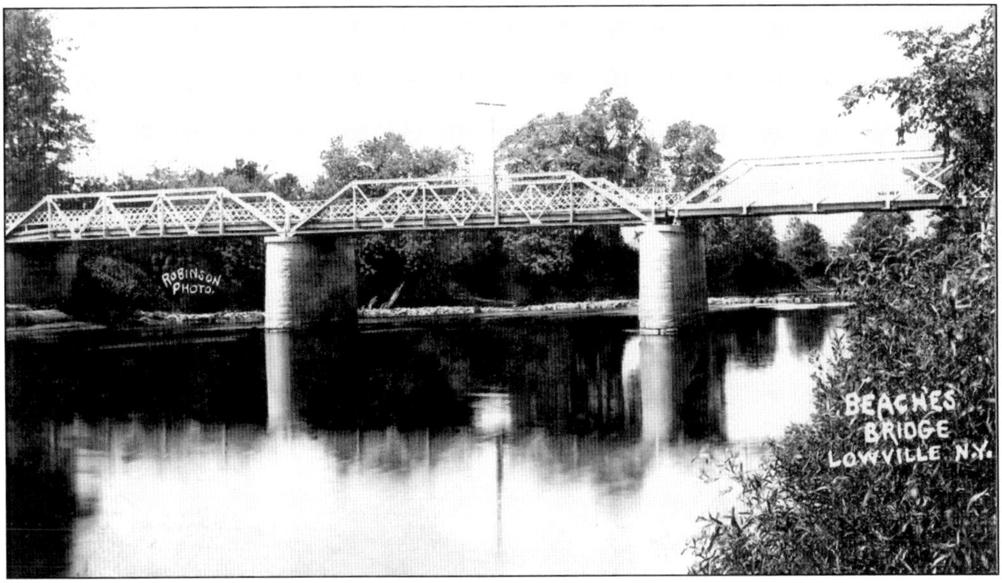


Beach's Bridge, built in 1851, spanned the Black River three miles north of Bush's Landing. The eastern section of this 185-foot-long bridge swung open to let steamers pass underneath. The bridge pictured here was built in the post-canal era using the earlier supports. (Courtesy of George Davis.)

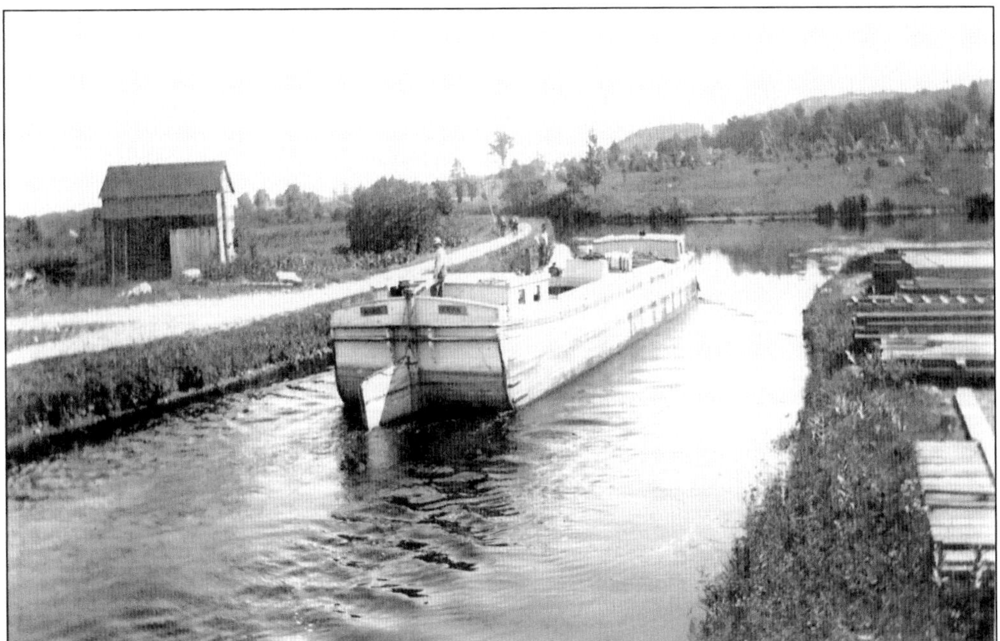

Capt. William Nichols is at the helm of the *Mabel Ervin*. This boat was built at the Van Amber Brothers' New Bremen boatyard, one of the largest on the canal. (Courtesy of Bob Hindman.)

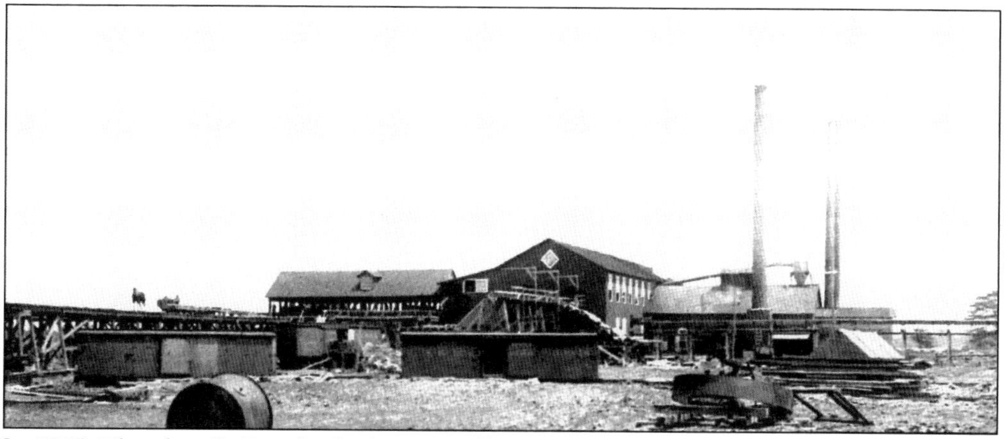

In 1885, Theodore B. Basselin built a sawmill at Castorland near the mouth of Beaver River. The mill employed 250 men and turned out 125,000 board feet a day for 14 years.

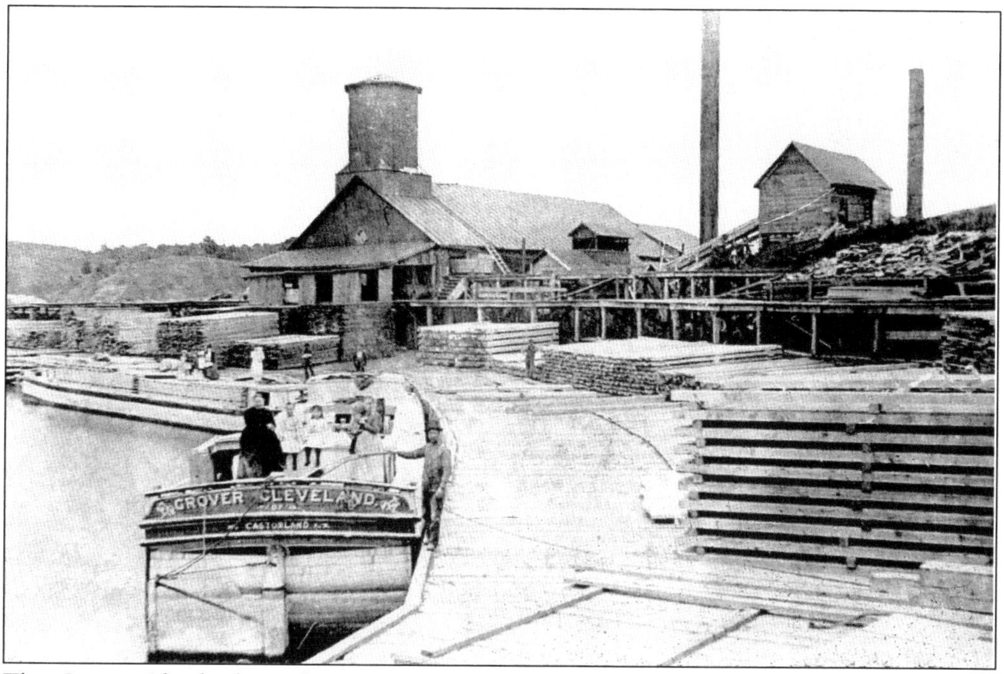


The *Grover Cleveland* is taking on scantling at Basselin's 10-acre lumberyard. A staunch Democrat, Basselin named a number of his canal boats after prominent figures in the party. (Courtesy of New York State Museum.)

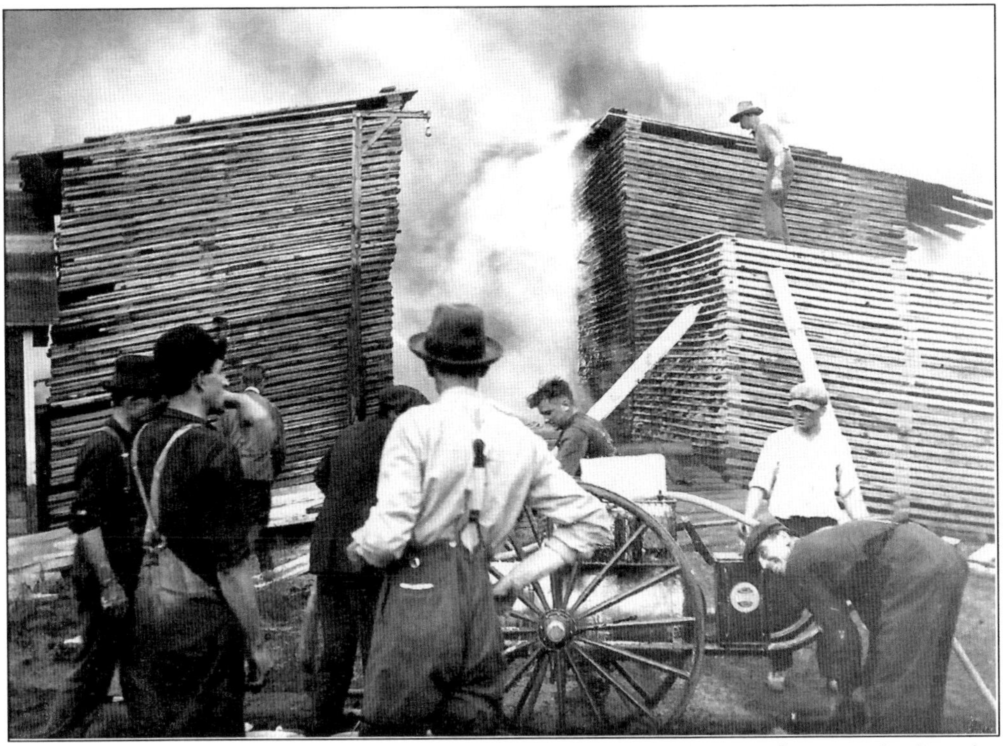

Fire was a constant threat at sawmills and lumberyards. Among the casualties was the Basselin mill, which burned down in 1899 despite the presence of 10 fire hydrants, 4 night watchmen, and fire whistles. When the second mill burned down in 1909, it was not rebuilt. But by that time, steamboat traffic on Black River had largely died out. The fire pictured here is at Brown's Tract Mill in Thendara. (Courtesy of Town of Webb Historical Association.)

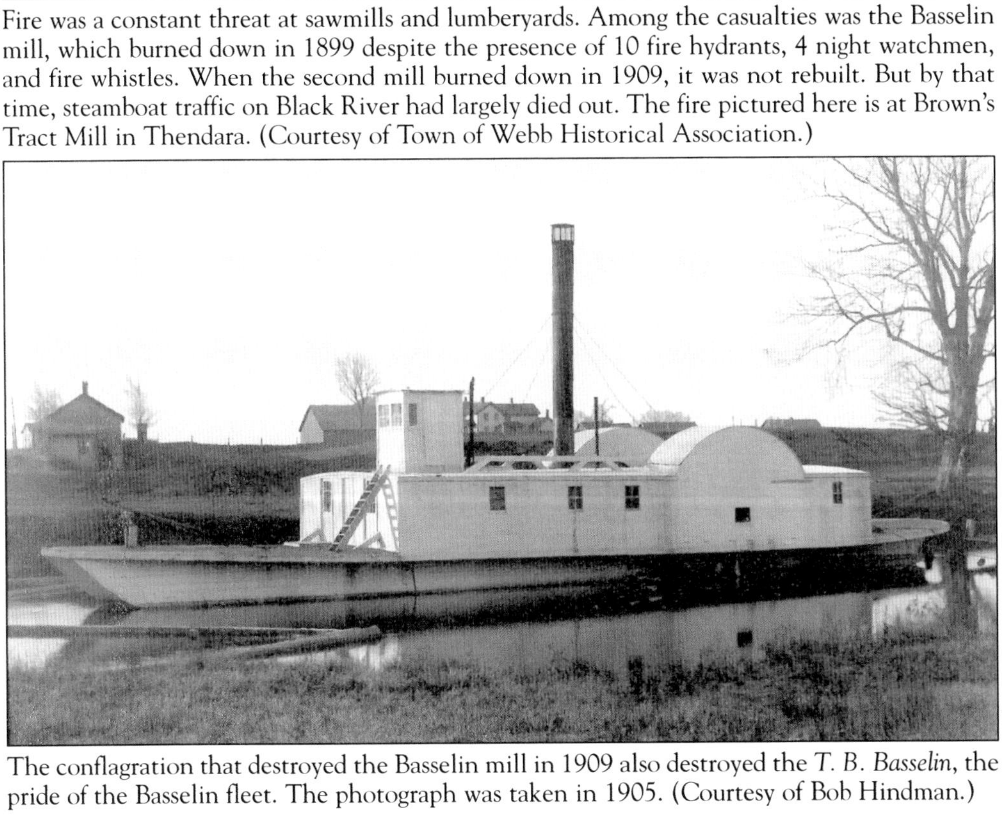

The conflagration that destroyed the Basselin mill in 1909 also destroyed the *T. B. Basselin*, the pride of the Basselin fleet. The photograph was taken in 1905. (Courtesy of Bob Hindman.)

Kohler and Wheelock ran a commission house in Carthage. Freight was shipped on the Carthage, Lowville, and New York Line, founded in 1858 by George Sweet, Nicholas Wagner, and Christopher Rhiner. The company operated six boats and had offices at the corner of Canal and West Streets in Carthage. Competition from the railroad forced the firm out of business in 1870. (Courtesy of New York State Museum.)

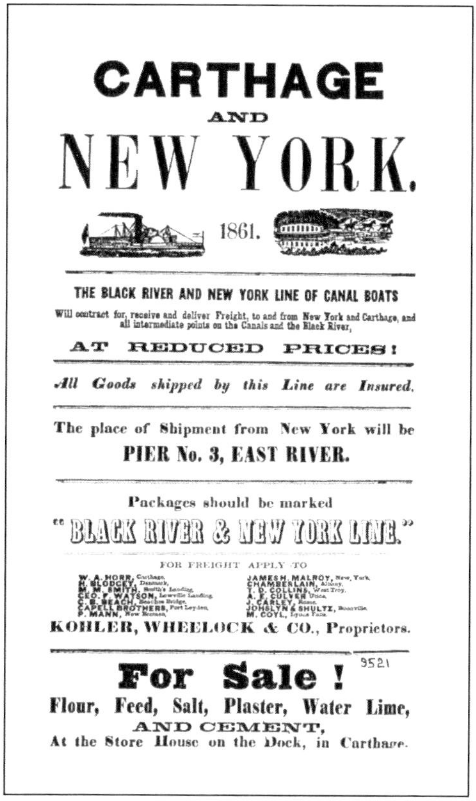

CARTHAGE
AND
NEW YORK.
1861.

THE BLACK RIVER AND NEW YORK LINE OF CANAL BOATS
Will contract for, receive and deliver Freight, to and from New York and Carthage, and all intermediate points on the Canals and the Black River,
AT REDUCED PRICES!
All Goods shipped by this Line are Insured.
The place of Shipment from New York will be
PIER No. 3, EAST RIVER.
Packages should be marked
"BLACK RIVER & NEW YORK LINE."
FOR FREIGHT APPLY TO
W. A. HORR, Carthage, JAMES H. MALROY, New York,
H. BLODGET, Denmark, CHAMBERLAIN, Albany,
M. M. SMITH, Smith's Landing, T. D. COLLINS, West Troy,
GEO. F. WATSON, Lowville Landing, A. E. CULVER Utica,
C. B. BEACH, Beaches Bridge, J. CARLEY, Rome,
CAPELL BROTHERS, Port Leyden, JOHSLYN & SHULTZ, Boonville,
P. MANN, New Bremen, M. COYL, Lyons Falls.
KOHLER, WHEELOCK & CO., Proprietors.
9521
For Sale!
Flour, Feed, Salt, Plaster, Water Lime,
AND CEMENT,
At the Store House on the Dock, in Carthage.

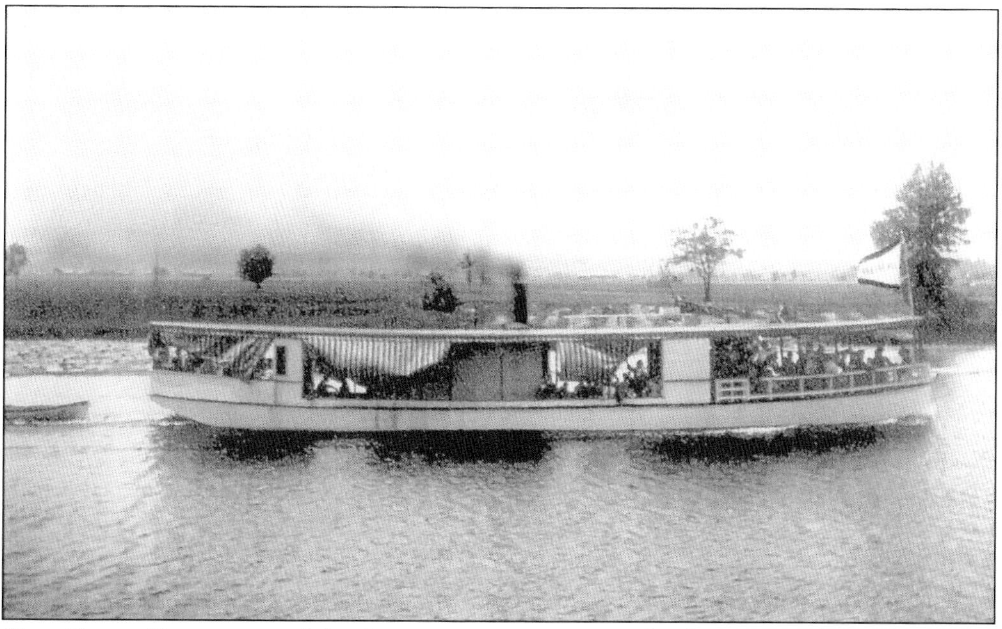

Capt. George Sweet operated the *Outterson*, an excursion yacht that conveyed passengers daily from Carthage to Cold Spring Park. This park was on the eastern bank of Black River opposite the mouth of Deer River. The steamer was named for James Andrew Outterson, who established numerous paper mills on Black River. (Courtesy of Lewis County Historical Society.)

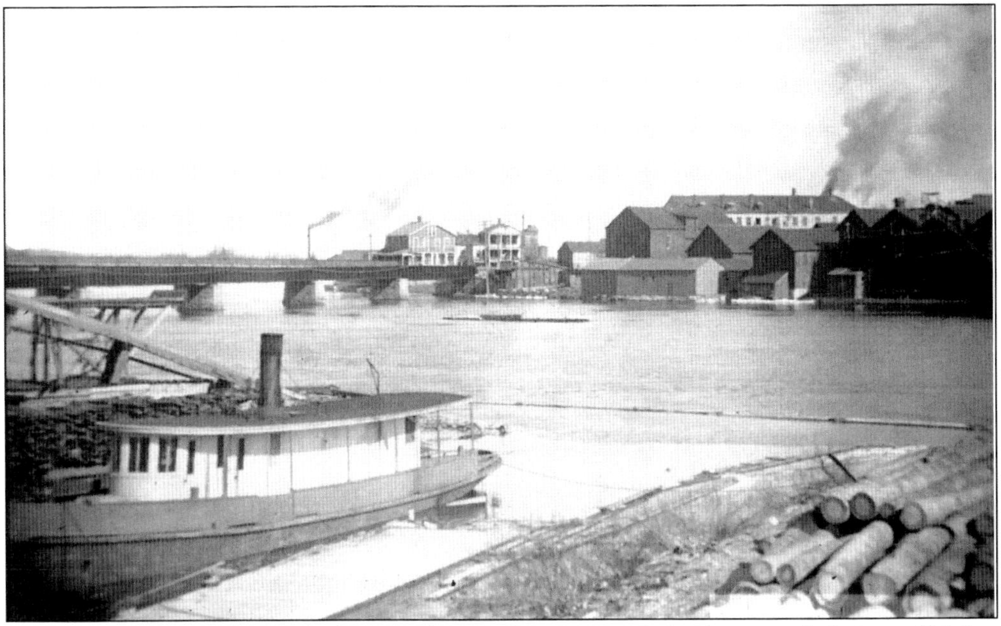

A small steam launch is docked at Carthage. The seven-arch bridge extended about 435 feet across Black River. (Courtesy of New York State Museum.)

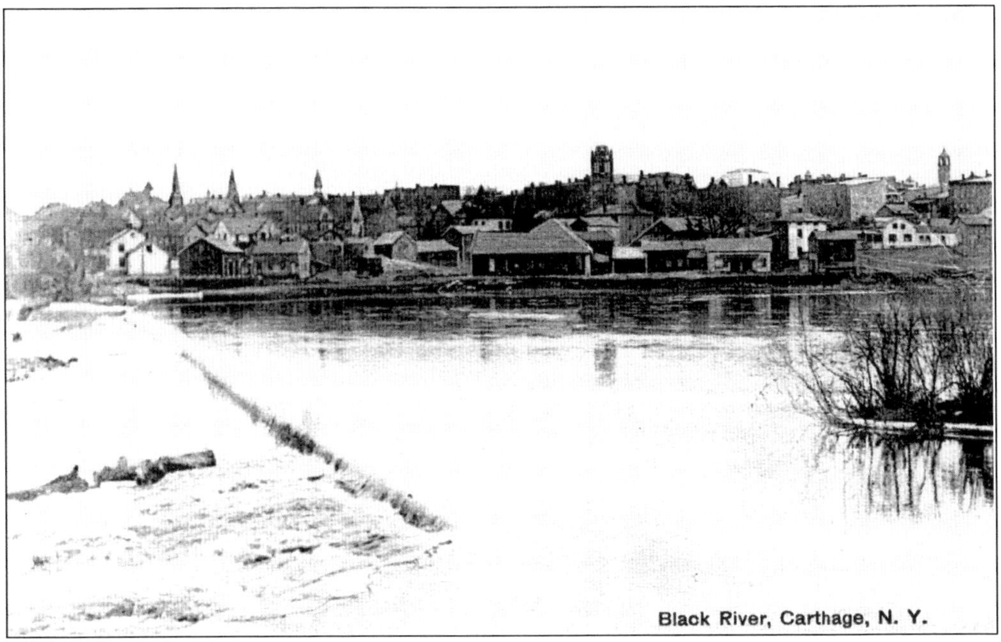


The state dam at Carthage marks the northern terminus of the Black River Canal. The dam made slack-water navigation possible as far Bush's Landing. (Courtesy of Erie Canal Museum.)

# Six

# LAST DAYS

From a strictly financial viewpoint, the Black River Canal was never a success. At the end of 1882, when the state ceased to collect tolls on all canals, the Black River Canal had yielded $301,000 in revenue, whereas the cost of operation and maintenance had reached more than $1.5 million. The deficit, in part, reflected the large cost of maintaining the Black River impoundment system. By 1900, however, tonnage on the canal had fallen drastically, and it had become clear that the canal had outlived its usefulness.

The canal suffered from inadequate funding, constant disrepair, the inability to accommodate large boats, and competition from the railroad. The principal reason for the decline in canal business, however, was the fact that the easy-to-reach softwood had largely disappeared. Timber—the mainstay of the canal—was no longer available in quantities large enough to sustain the canal as a viable enterprise. Whereas forest products accounted for 94 percent of canal tonnage in 1893, they accounted for only 38 percent in 1903.

A temporary reprieve for the canal came with the construction of the Barge Canal system between 1903 and 1918. Massive quantities of limestone from Sugar River and high-quality sand from Boonville and Frenchville were shipped to construct Delta Dam on the Mohawk River 5.5 miles north of the Erie Canal. These shipments made 1910 the busiest year of the canal. But once the new canal was completed, traffic slowed to a trickle, and the Black River Canal was finally abandoned by the state in 1922. The last boat, the state scow, left Boonville on July 14, 1924.

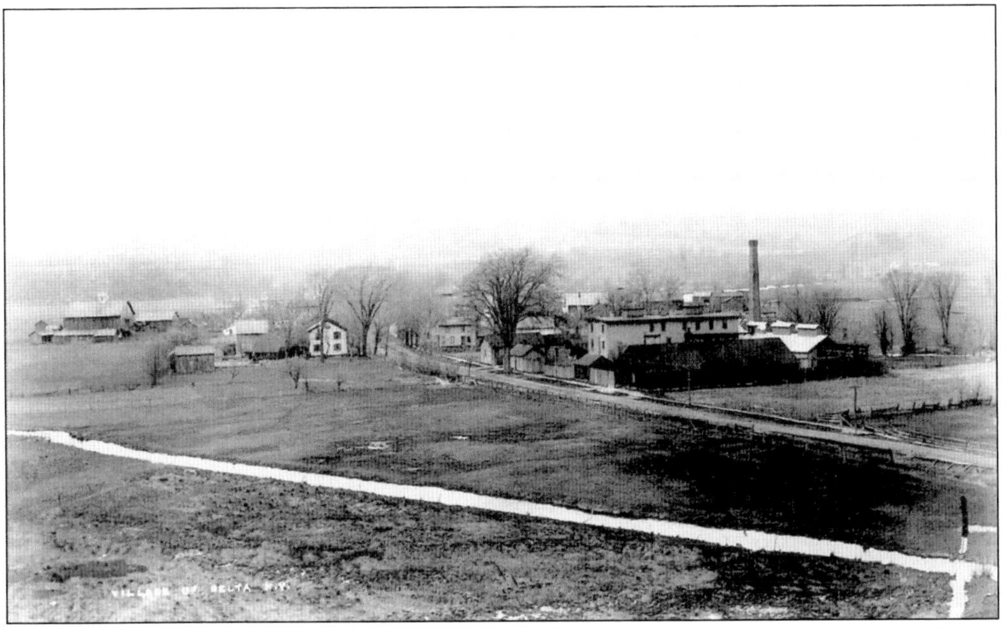

The Black River Canal bypassed the village of Delta. In 1860, however, a 1.25-mile feeder connecting the Mohawk River with the Black River Canal was completed.

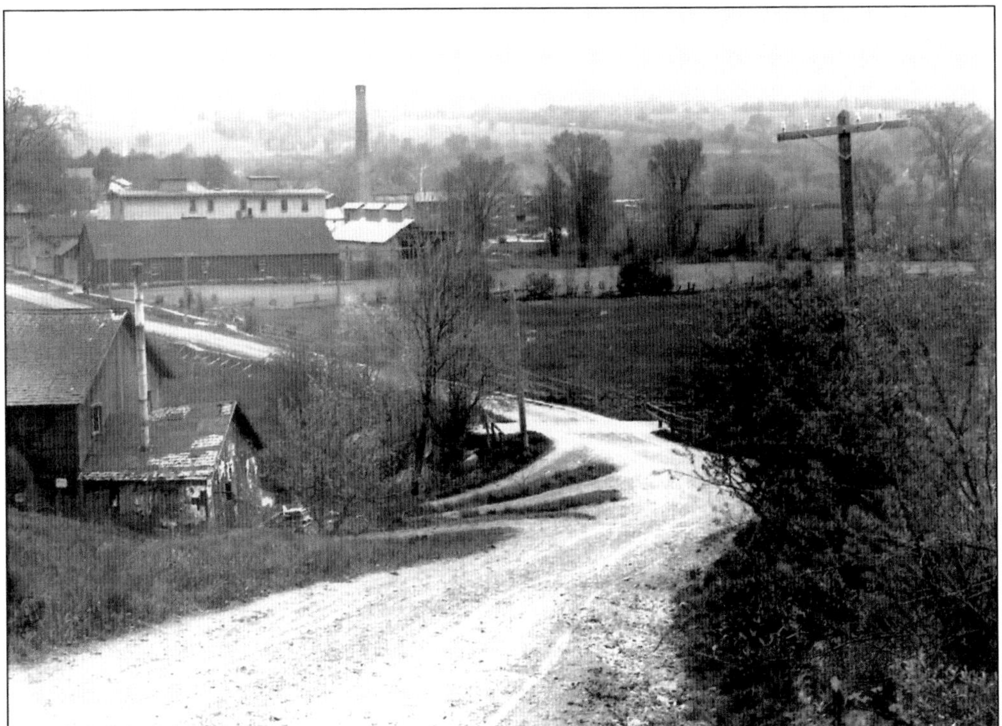

C. Frank Floyd and George B. Olney built a canning factory at Delta in 1884. Within a few years, the factory was turning out 400,000 cans of vegetables. Tin for cans was imported on the Delta Feeder. (From Rome Historical Society's Photographic Collection.)

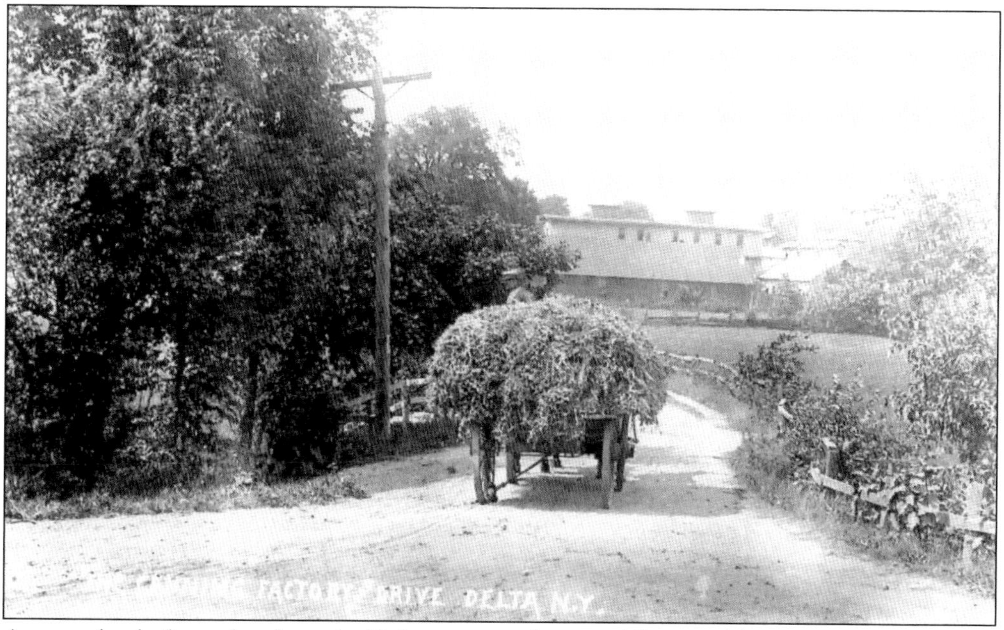


A wagonload of peas heads to the Delta canning factory. All vegetables canned at the factory were locally grown.

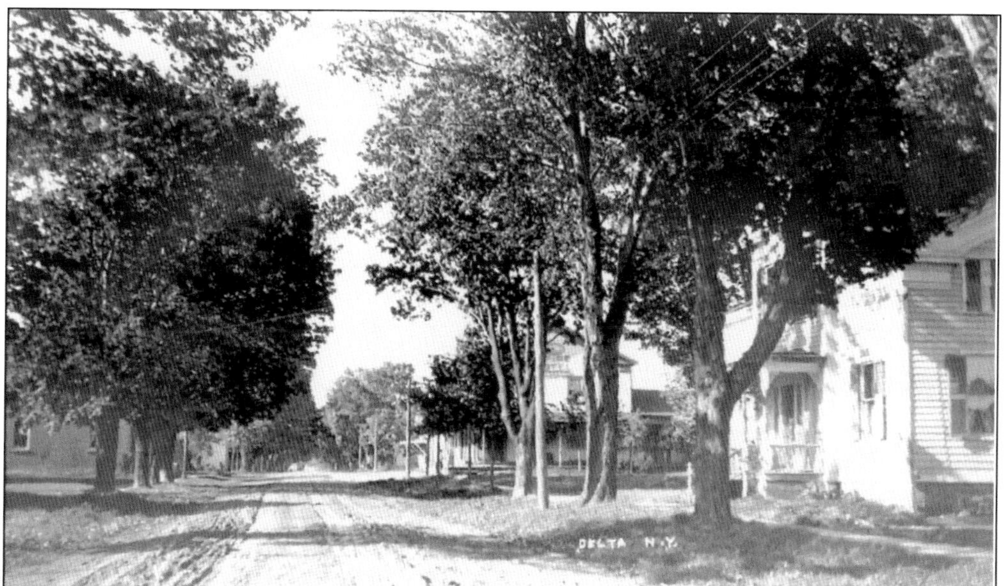


In 1912, the Mohawk River was dammed to create Delta Lake, a reservoir for the new Barge Canal. To make space for the reservoir, some 250 buildings in the village of Delta had to be moved. Engineers had considered flooding Westernville as well, but political opposition defeated the plan. (From Rome Historical Society's Photographic Collection.)

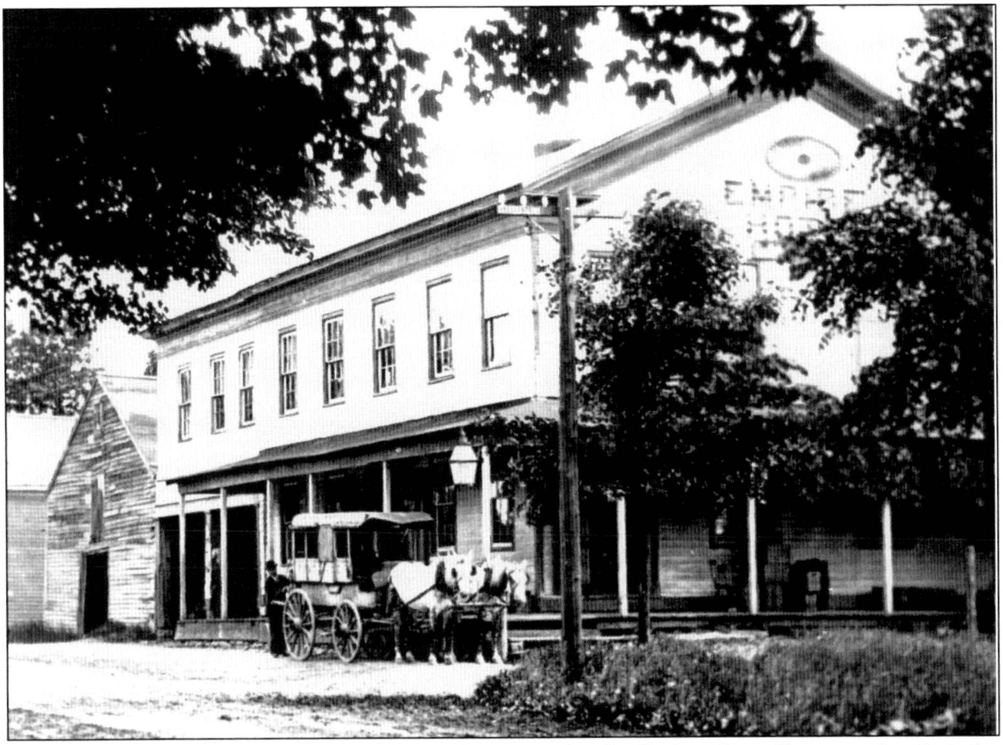

Eliakim Elmer built the Empire Hotel in Delta around 1854. Before the village was flooded, the Empire was dismantled and moved to Westernville, where it eventually became the town hall. (From Rome Historical Society's Photographic Collection.)

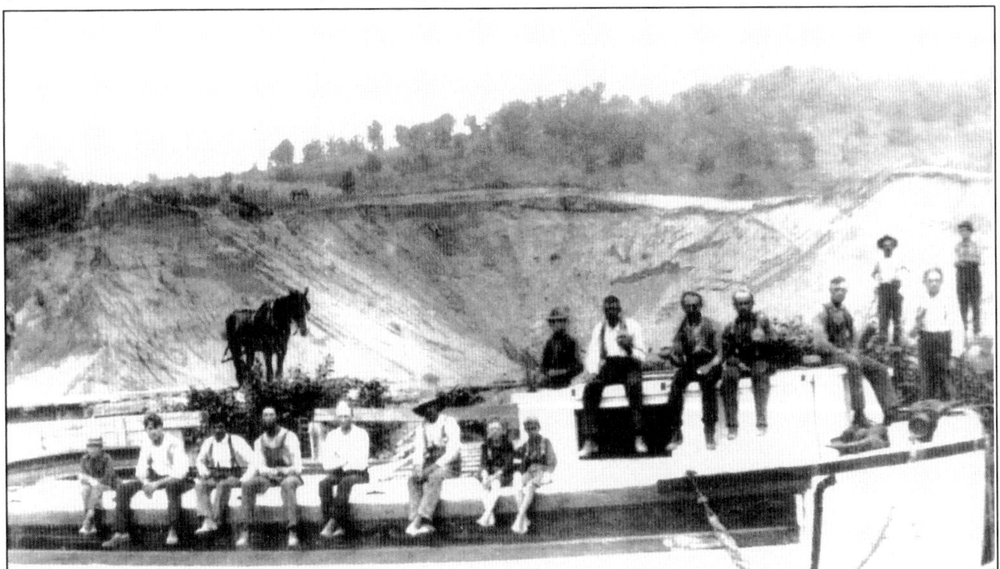

The Black River Canal ran past several sandpits in Boonville. Especially well suited for masonry, this sand was shipped to the Paragon Plaster Company of Syracuse in large quantities. In the 1950s, Boonville sand helped construct the New York State Thruway.

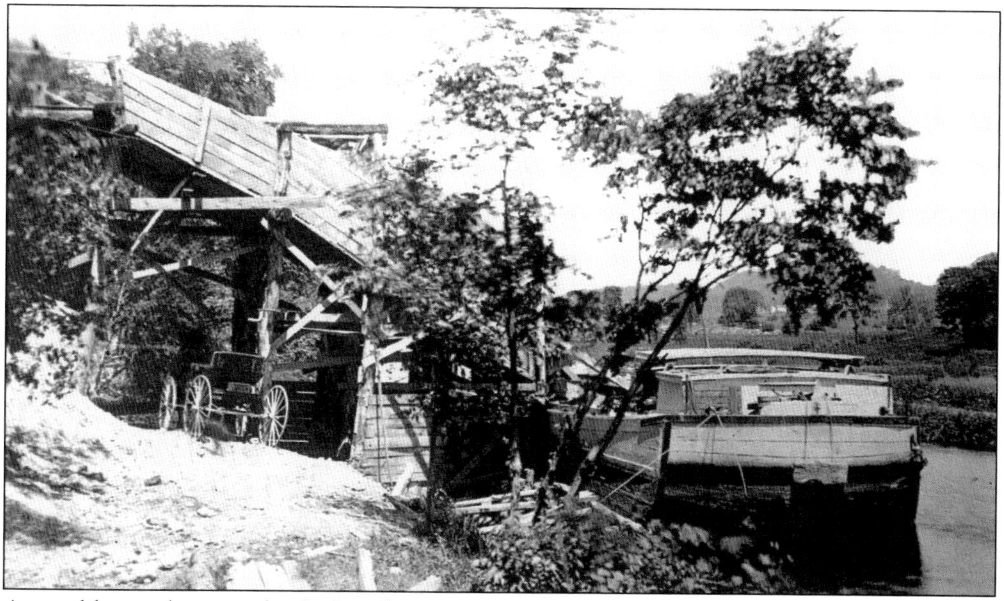

A canal boat takes on a load of sand at Hough's landing in Boonville. (Courtesy of New York State Museum.)

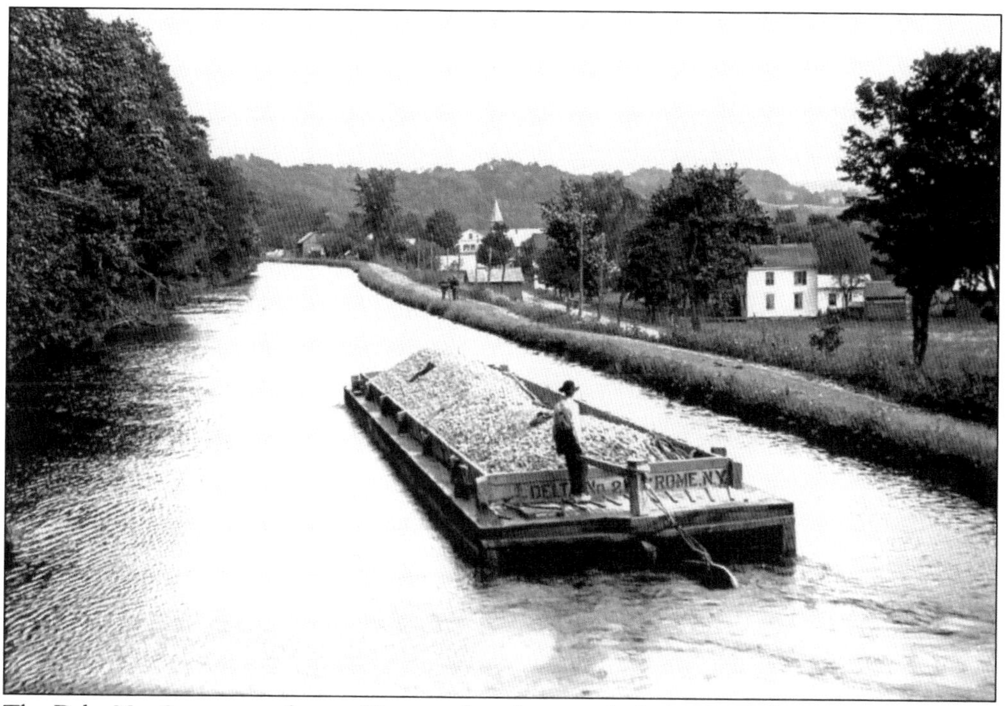


The *Delta No. 2* was one of some 30 scows that the state built for the Delta project. The scene is at North Western. (Todd Weseloh Collection.)

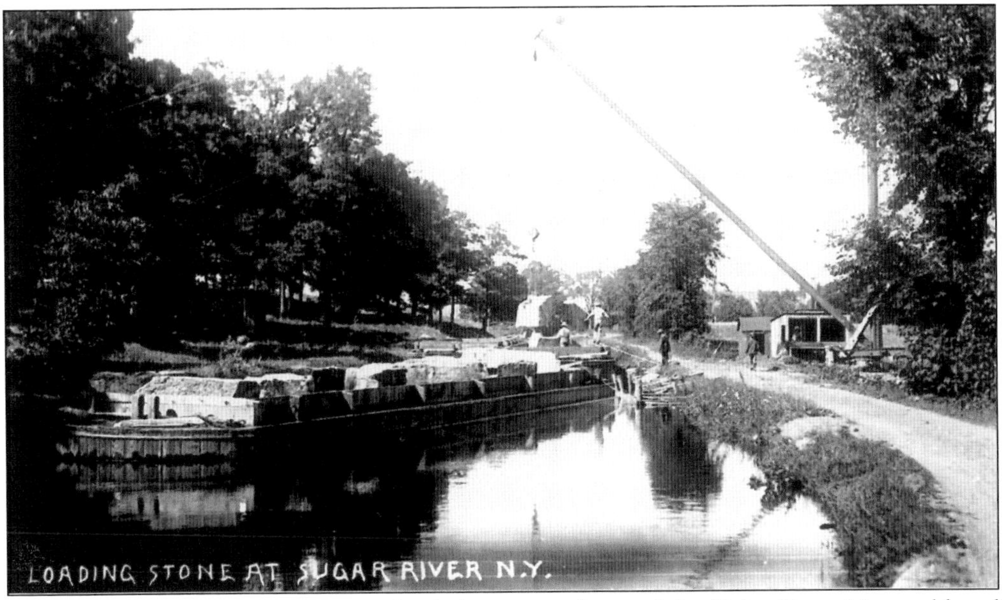


At the Sugar River quarry, a crane loads limestone destined for Delta Dam. Many old and retired boats were pressed back into service for this construction project. In 1910, some 40 boats kept busy transporting materials to Delta.

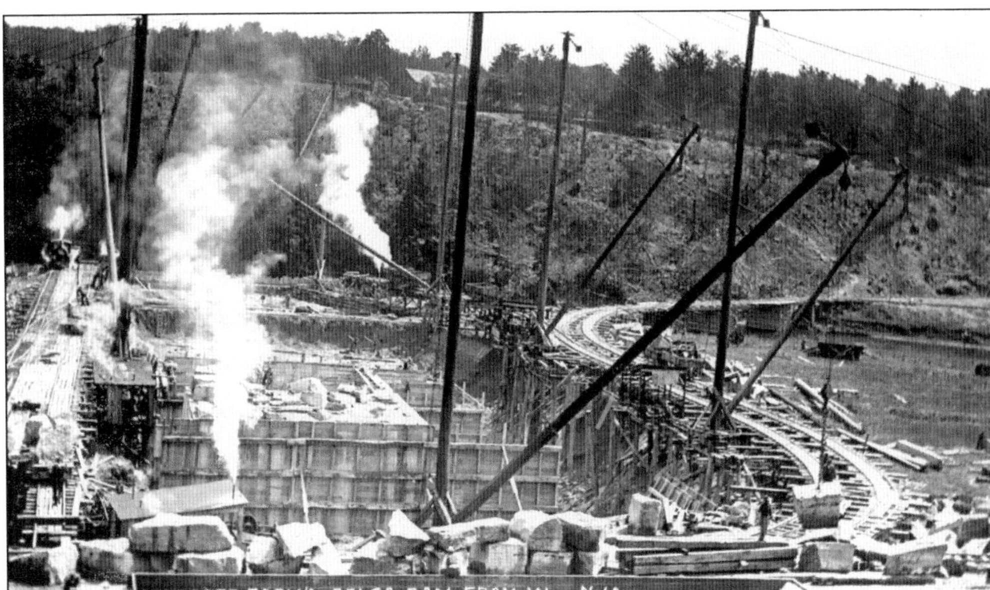


The new dam required 90,000 cubic yards of masonry, 45 percent of which was dimension limestone from Sugar River. (Courtesy of Mary Centro.)

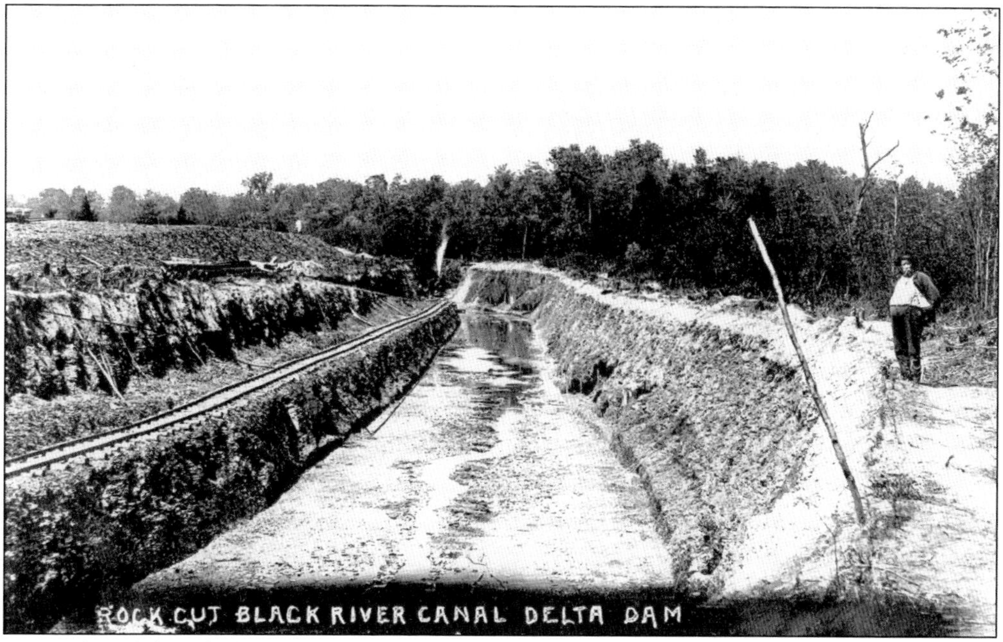


Two miles of the Black River Canal were relocated near Lock 13. A 70-ton steam shovel was used to excavate a 2,500-foot canal bed in the shale along the eastern bank of Lake Delta.

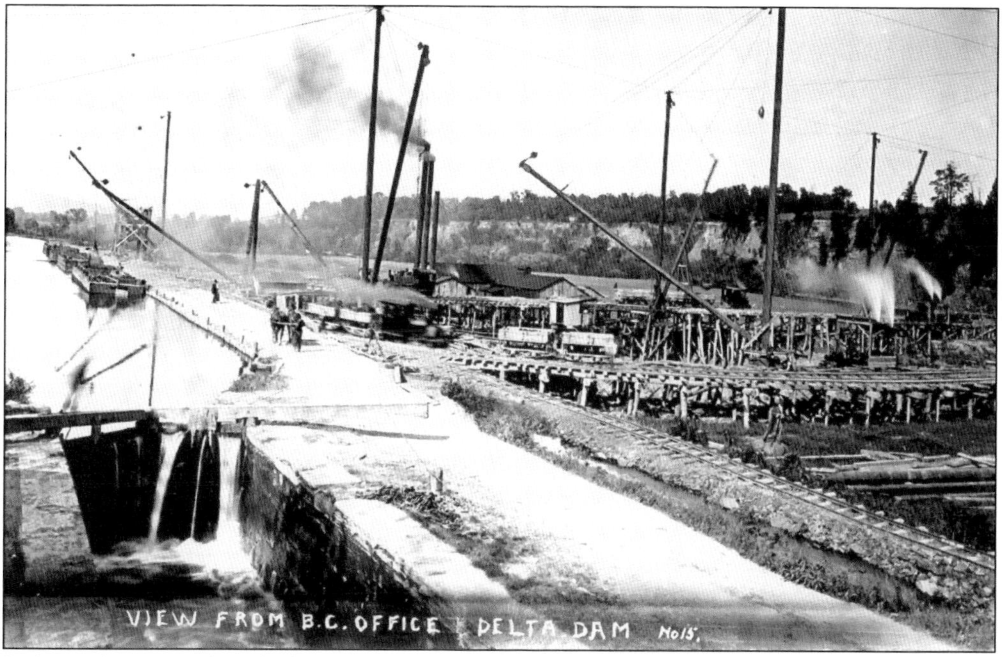


Before relocation, the Black River Canal ran along the western side of the Mohawk River at Delta, as pictured here. (Courtesy of Mary Centro.)

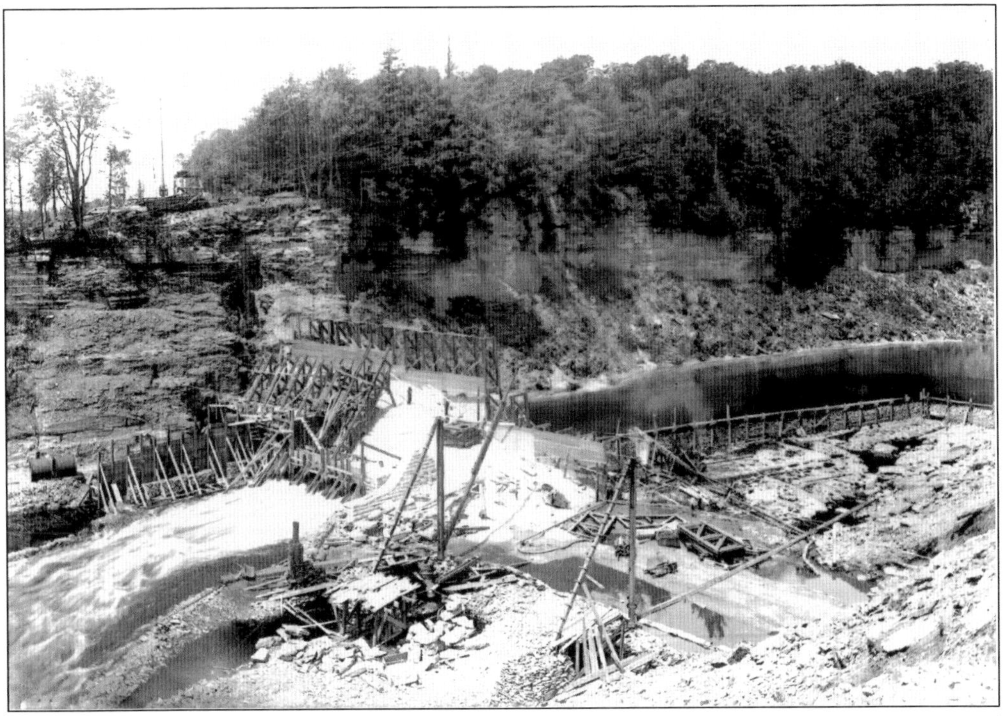

Delta Dam was built in 30-foot sections separated by expansion joints. The dam is situated in what had been the outlet to Lake Iroquois, a massive glacial lake to the north. This outlet carved tall palisades that served as walls for the dam.

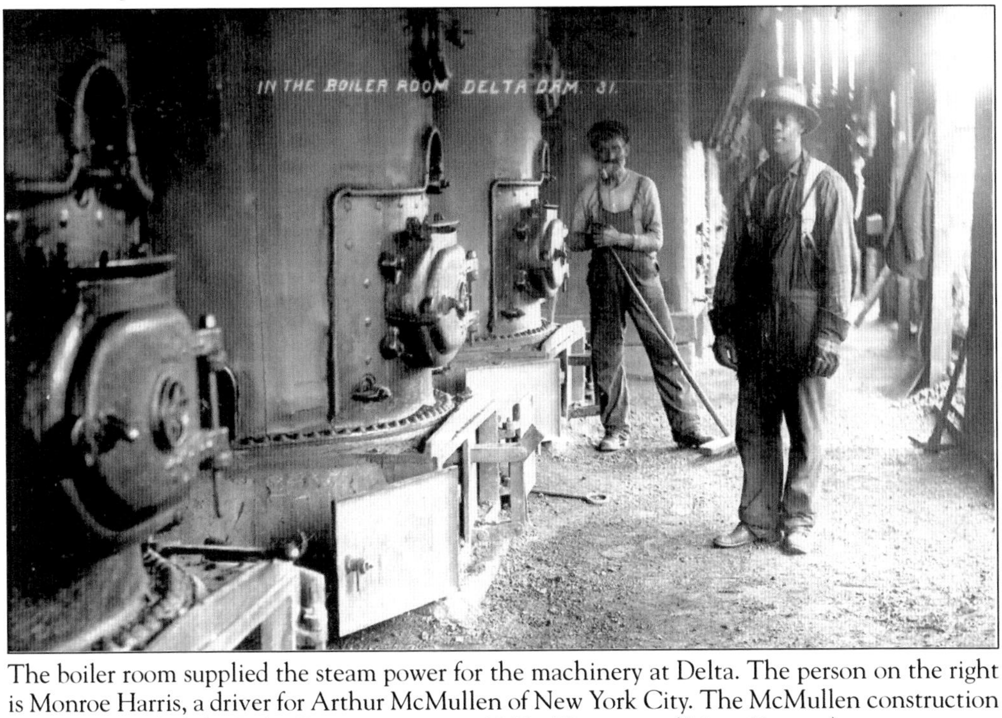


The boiler room supplied the steam power for the machinery at Delta. The person on the right is Monroe Harris, a driver for Arthur McMullen of New York City. The McMullen construction firm was awarded the Delta Dam contract in 1908. (Courtesy of Mary Centro.)

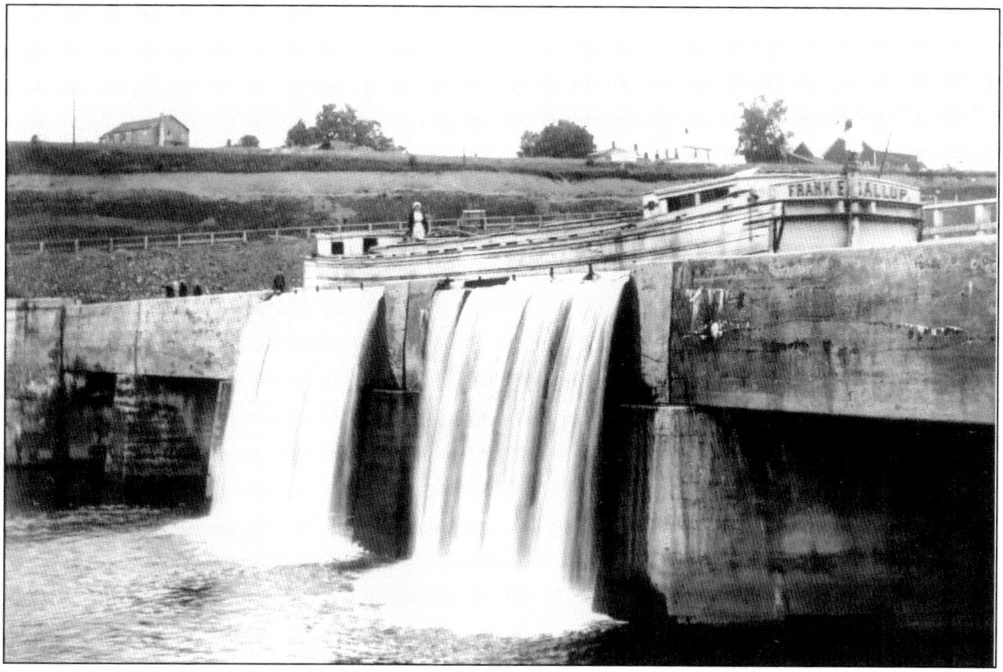


The *Frank E. Gallup*, captained by Jesse Cronk, crosses the aqueduct at Delta while carrying a load of sand from the Gallup-Draper pit in Boonville. (From Rome Historical Society's Photographic Collection.)

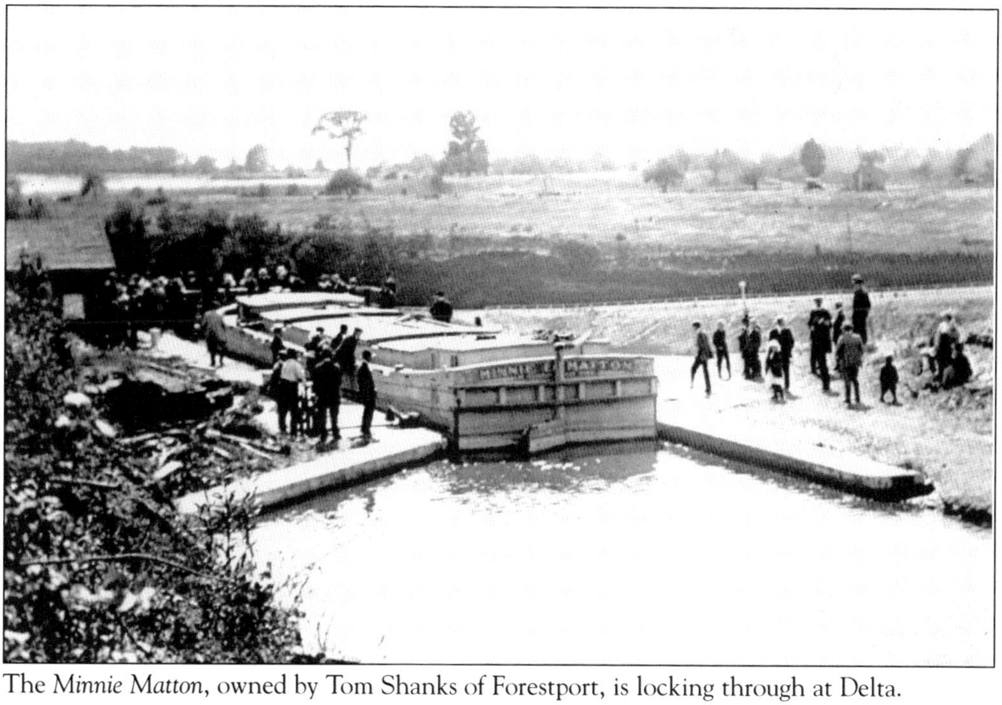

The *Minnie Matton*, owned by Tom Shanks of Forestport, is locking through at Delta.

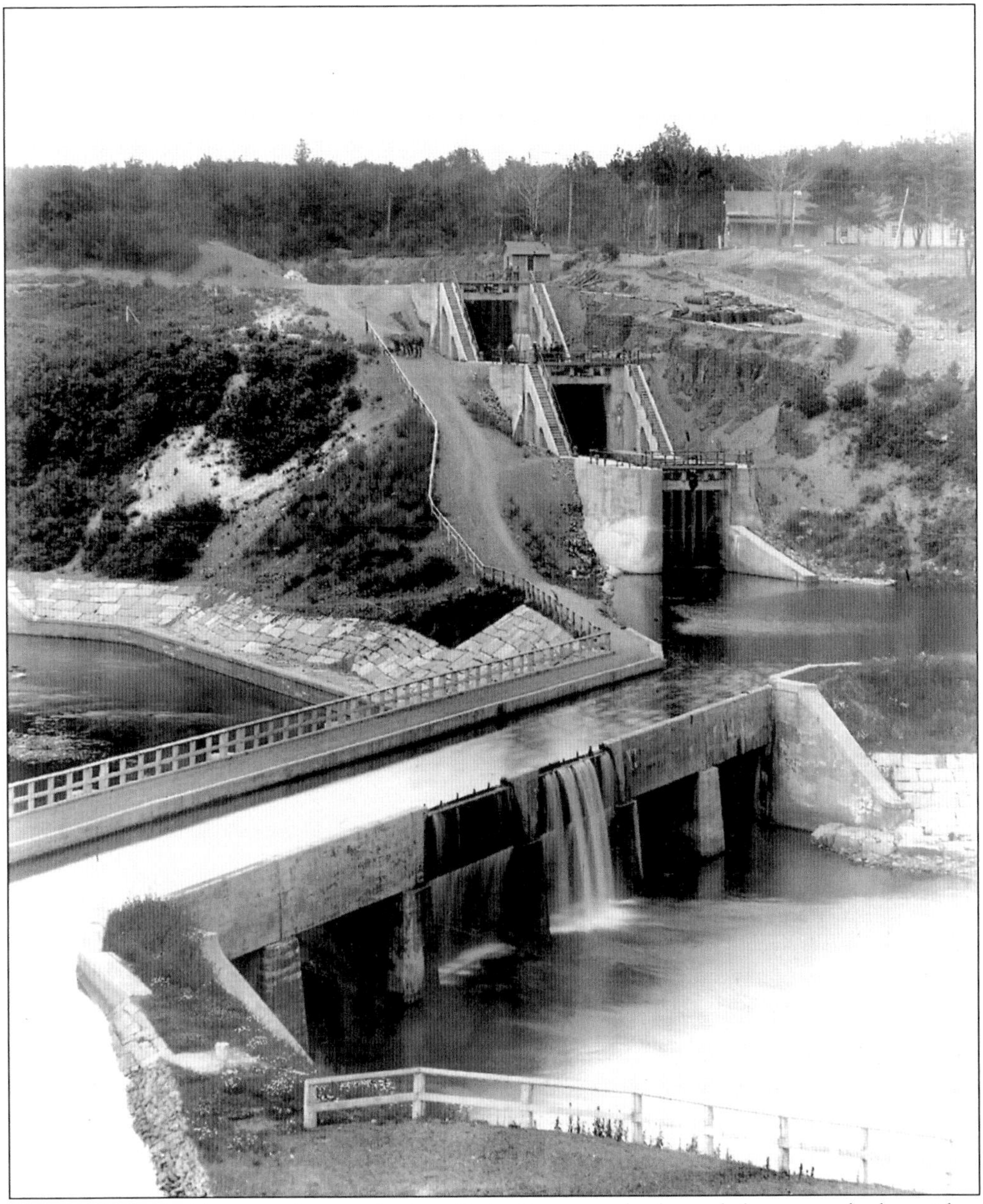

Locks 7 through 13 were submerged when Delta Dam was built, and three new locks, with a lift of 62 feet, were built. A new aqueduct, extending the canal across the Mohawk, was also constructed. The new locks and aqueduct were in use for only about a decade. (From Rome Historical Society's Photographic Collection.)

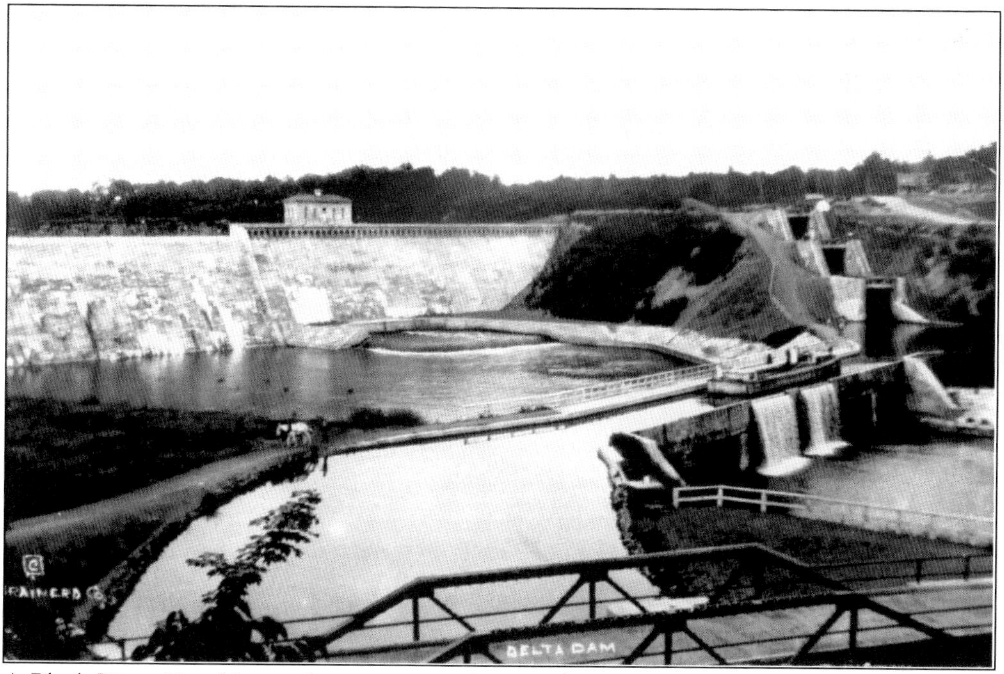


A Black River Canal boat advances across the aqueduct at Delta. The dam is 1,100 feet long and 100 feet high with a 300-foot spillway. A concrete apron was built 10 feet below the river to form a pool that would cushion the falling water.

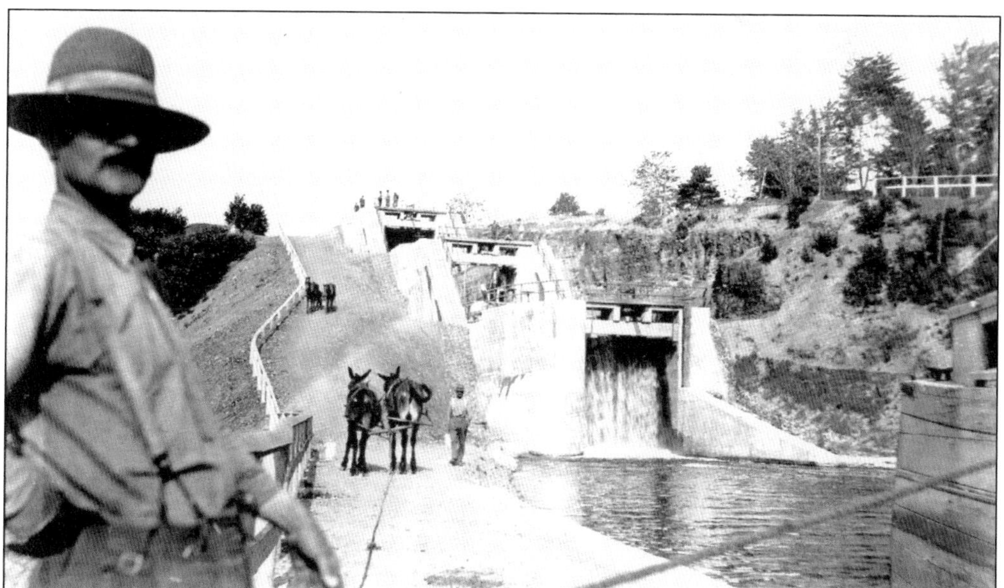

A northbound driver waits for southbound boats to lock through at Delta. In the early 1900s, drivers typically earned $1 a day plus board. On the Barge Canal, boats were no longer towed by draft animals but were instead either self-propelled or towed by tugs.

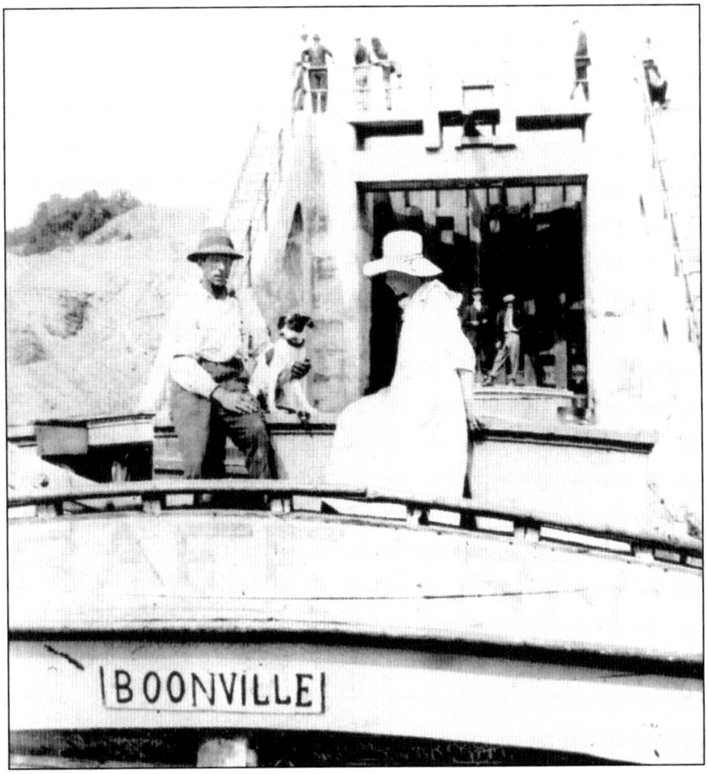


Port Leyden residents Arthur and Celia Lane pose with their dog on the *Charles Traffarn*, a Boonville boat. Such boats appeared miniscule next to the 3,000-ton boats operating on the Barge Canal.

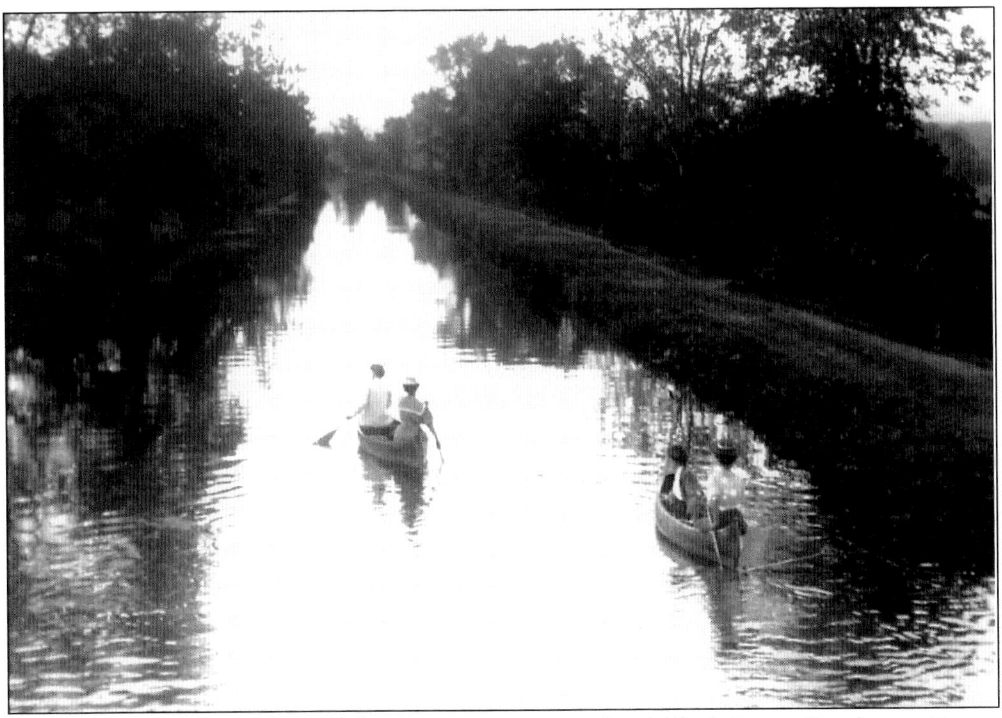

Fashionably dressed women paddle their canoes on a placid Black River Canal near Rome. (From Rome Historical Society's Photographic Collection.)

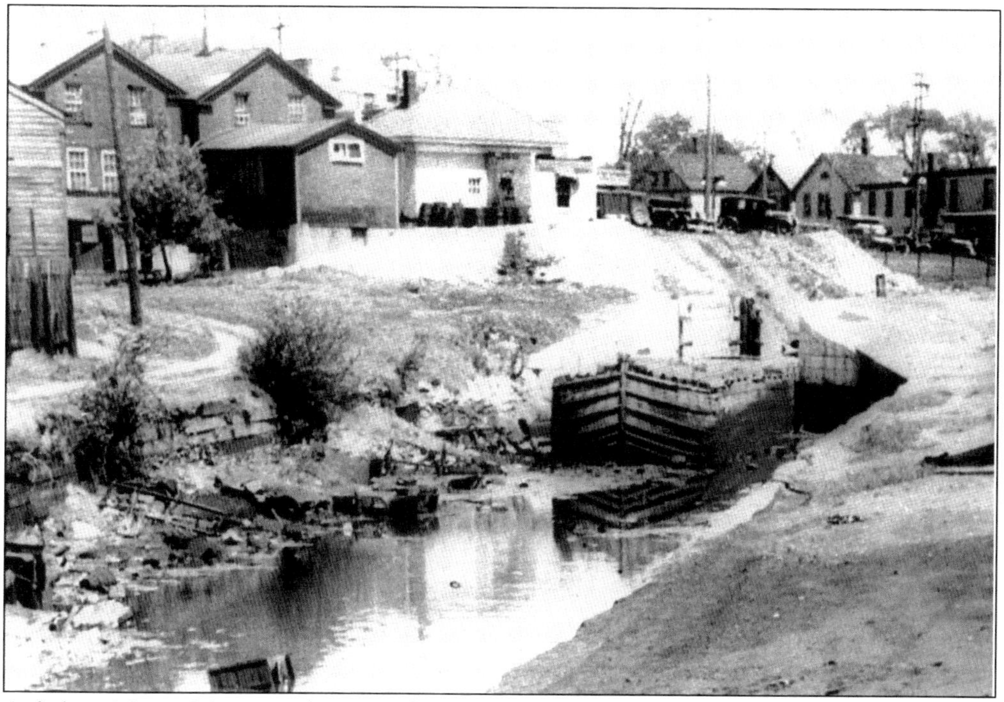

A forlorn *Minnie Matton* endures another winter at Lock 1 in Rome. (Courtesy of New York State Museum.)

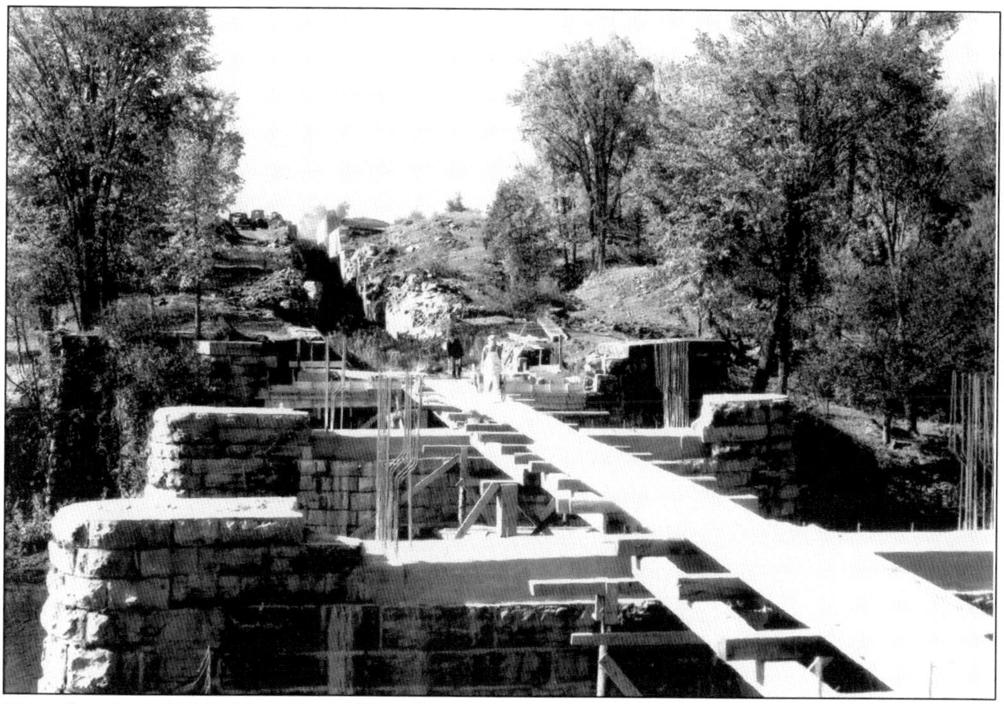

Completed in 1935, New York State Route 12 now crosses Sugar River at the site of the canal aqueduct. A few locks between Boonville and Lyons Falls were destroyed when the new highway was built, but most remain intact and in relatively good condition.

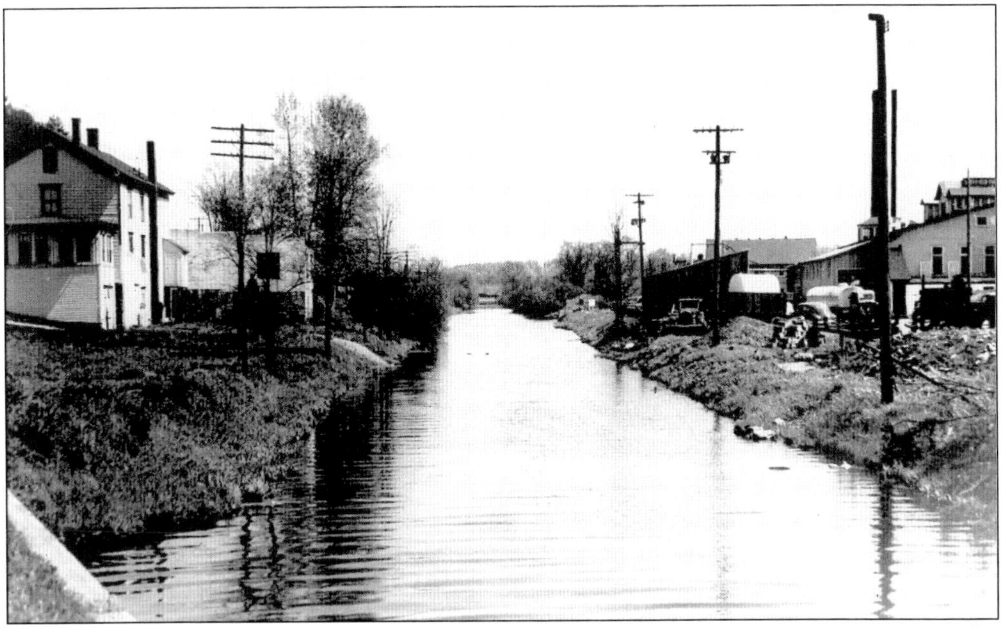

This view of the Black River Canal is looking south from the Main Street Bridge in Boonville. Fisher Street, on the left, disappeared when Route 12 was expanded in 1959. The canal itself was built on the site of the former South Street. In recent years, the right bank of the canal has been converted into a walkway and park. (Courtesy of New York State Museum.)

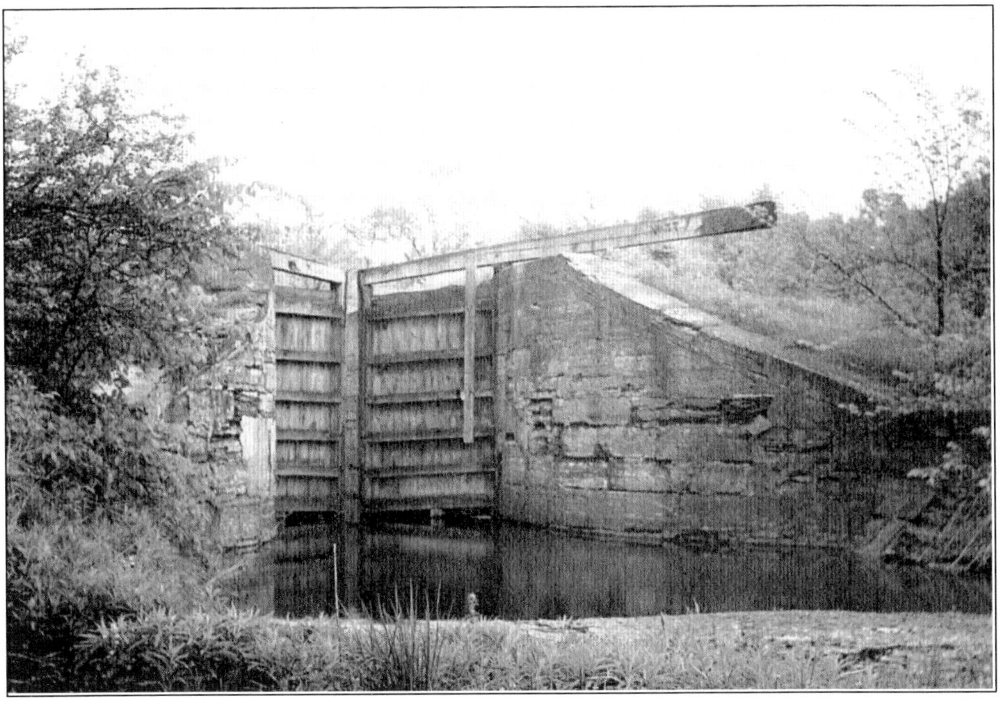

When this picture was taken in the 1940s, the wooden gates in Lock 37 were more or less intact. The many surviving locks of the Black River Canal make the former waterway a prime candidate for inclusion in the National Heritage Corridor. (Todd Weseloh Collection.)